RECENT TRENDS IN BIOTECHNOLOGY

FORENSIC SCIENCE

NEW DEVELOPMENTS, PERSPECTIVES AND ADVANCED TECHNOLOGIES

RECENT TRENDS IN BIOTECHNOLOGY

Additional books in this series can be found on Nova's website under the Series tab.

Additional e-books in this series can be found on Nova's website under the e-book tab.

RECENT TRENDS IN BIOTECHNOLOGY

FORENSIC SCIENCE

NEW DEVELOPMENTS, PERSPECTIVES AND ADVANCED TECHNOLOGIES

JOHANNA BREWER
EDITOR

nova publishers
New York

Library of Congress Cataloging-in-Publication Data

Forensic science : new developments, perspectives and advanced technologies / Johanna Brewer, editor.
pages cm. -- (Recent trends in biotechnology)
Includes bibliographical references and index.
ISBN 978-1-63483-086-7 (hardcover)
1. Medical jurisprudence. I. Brewer, Johanna.
RA1051.F667 2015
614'.1--dc23
2015019753

Published by Nova Science Publishers, Inc. † New York

CONTENTS

PREFACE

This book explores recent developments in forensic science research, including invisible radiation imaging, providing important insights into evidence normally beyond the visual experience of investigators. Additionally, establishing the interval between the time of death and when a body is found is one of the most complex questions to be answered by forensic scientists. The second chapter examines new approaches in postmortem interval (PMI) estimation. Finally, in forensic medicine, the diagnosis of a corpse immersed in water in which a differentiation must be made between death from drowning or dead on entering the water, is made mainly using the diatom test by acid digestion. The authors assess the 16S rDNA gene of picoplankton from tissues. The results verified that the detection of phytoplanton DNA in the liver and kidney is the most important evidence for the diagnosis of death from drowning.

Chapter 1 – Invisible radiation imaging provides important insights into evidence normally beyond the visual experience of investigators. Reflected ultraviolet (UV) photography has classically been used for recording bite marks, bruises, car panel damage and fingerprints. The recent application of UV digital imaging potentially provides many advantages for forensic investigation as images can be viewed in real time at a crime scene, potentially enabling efficient collection of critical evidence that previously went unseen. However, since UV imaging collects data that is beyond our normal frame of reference for interpreting results, it is important that robust methodologies can be applied to quantify relative reflectance from different elements of a potential crime scene. We discuss the dynamics of the non-linear relationships between reflected radiation and the response of commercial grade image sensors that are typical in forensic practice, and how the implementation of image processing algorithms based on non-linear functions enables the recovery of robust linearized data for the precise quantification of reflectance in a scene. We demonstrate the application of this process with both a typical (fingerprint) and novel (material identification based on its reflective properties) problem for forensic imaging, and discuss how this linearized process will allow for the accurate documentation of reflected UV imaging as evidence in court proceedings.

Chapter 2 – Establishing the interval between the time of death and when a body is found is one of the most complex questions to be answered by forensic scientists. Postmortem interval (PMI) estimation is absolutely crucial in forensic science for the reconstruction of perimortem events, leading to possible criminal and civil repercussions. In order to estimate accurately the time since death, it is essential to have adequate knowledge in decomposition process and the impact of the different environmental conditions in that process. The postmortem changes useful for estimating time since death span from different processes: physical (like body cooling and hypostasis); metabolic (supravital reactions); physic-chemical (*rigor mortis*); bacterial (putrefaction); autolysis (loss of selective membrane permeability, diffusion) and insect activity. However, as time passes, taphonomic factors accumulate their effects on the corpse, being less precise and challenging the estimation of postmortem interval. As a result, PMI is really difficult to be accurately determined, so that forensic researchers are in need of modern tools to assist in PMI estimation. Scientific efforts should be made to replace the traditional methods of estimation the time since death by those that calculate this parameter providing mean values and confidence limits, including the precision and accuracy of calculation. Continuing research is developing towards this objective. Numerous methods have been proposed for the determination of the time since death mainly by chemical means, like the measure of volatile fatty acids in the soil solution, as well as the analysis of amino acids, neurotransmitters and decompositional by-products, leading to the development of the new field in PMI estimation called thanatochemistry. Respect to molecular biology approaches, RNA appears as very promising target for the estimation of PMI. Its conspicuity across tissues with different degrees of exposure to diverse environmental factors and decay rates points out to more precise calibration of the PMI, by estimating and comparing its degradation rates and state across different organs of the subject. Diverse studies of post-mortem RNA degradation have shown that RNA's stability depends on the tissue and its presence also in the long-lived hard tissues of bone and teeth conducts to the possibility of extending the potential PMI estimates to longer time spans. In this line of research, it was recently demonstrated the determination of early PMI (2-6 hours after death) by analyzing the expression of two cell death proteins, FasL and PTEN, opens the door to a new approach in PMI estimation. New growing field towards PMI estimation is the study of bacterial community changes during body decomposition. After cessation of life occurs, human immune system interrupts its function, leading to a massive release of nutrients from autolytic human cells. As a consequence of that, microbes proliferate after human death. The composition and abundance of microbial communities change as function of time and temperature and this factor could be an accurate indicator of the elapse time-since-death. This emerging line of research joins efforts with the previous new approaches for PMI estimation.

Chapter 3 – In forensic medicine, the diagnosis of a corpse immersed in water that is differentiation between death from drowning or dead on entering the water, is made mainly using the diatom test by acid digestion. However, the diatom test by acid digesting is technically complicated and requires a larger quantity of sample. Moreover, the procedure is hazardous due to the use of strong acid and is time consuming. Alternative methods for the diagnosis of drowning by detecting plankton genes living in water using molecular biological technique have been reported. By performing one PCR amplification using various organs of a corpse immersed in water, this method is anticipated to permit rapid and accurate diagnosis of death from drowning by identifying the genes of phytoplankton that had entered solid organs via blood circulation. However, the currently used PCR method for the detection of phytoplankton requires DNA purification, necessitating treatment of tissues as well as differentiation between sea water and fresh water phytoplankton. Thus this method has the disadvantages of being labor intensive and having the risk of contamination. To overcome these issues, first we targeted synechococcus sp (Bacteria; Cyanobacteria; Oscillatoriophycideae; Chroococcales), a picoplankton found in both sea water and fresh water, and designed primers specific for its 16S ribosomal RNA genes (16S rDNA). We succeeded to find that phytoplankton DNA used the quantity of the organization from the organ of the drowned body using a direct PCR method without DNA extraction to check the death from being drowned from autopsy findings and a diatom test by acid digestion and an autopsy case to be diagnosed. Furthermore, we quantitatively assayed the 16S rDNA gene of picoplankton from tissues of drowned rabbits and non-drowned rabbits immersed in water after death. The research was to investigate the quantities of picoplankton DNA in lung, liver, kidney tissues and blood in drowned and non-drowned rabbits, and the sensitivity of detection of picoplankton DNA by direct PCR for the diagnosis of death from drowning. The results verified that the detection of phytoplankton DNA in the liver and kidney is the most important evidence for the diagnosis of death from drowning.

In: Forensic Science
Editor: Johanna Brewer

ISBN: 978-1-63483-086-7

Chapter 1

UV Digital Imaging: New Perspectives for Quantitative Data Analysis in Forensics

Jair E. Garcia and Adrian Dyer*
School of Media and Communication, RMIT University, Melbourne, Australia

Abstract

Invisible radiation imaging provides important insights into evidence normally beyond the visual experience of investigators. Reflected ultraviolet (UV) photography has classically been used for recording bite marks, bruises, car panel damage and fingerprints. The recent application of UV digital imaging potentially provides many advantages for forensic investigation as images can be viewed in real time at a crime scene, potentially enabling efficient collection of critical evidence that previously went unseen. However, since UV imaging collects data that is beyond our normal frame of reference for interpreting results, it is important that robust methodologies can be applied to quantify relative reflectance from different elements of a potential crime scene. We discuss the dynamics of the non-linear relationships between reflected radiation and the response of commercial grade image sensors that are typical in forensic practice, and how the implementation of image processing algorithms based on non-linear functions enables the recovery of robust

* Email: adrian.dyer@rmit.edu.au

linearized data for the precise quantification of reflectance in a scene. We demonstrate the application of this process with both a typical (fingerprint) and novel (material identification based on its reflective properties) problem for forensic imaging, and discuss how this linearized process will allow for the accurate documentation of reflected UV imaging as evidence in court proceedings.

INTRODUCTION

The ability to capture evidence with a camera has been a fundamental building block for how law enforcement investigators can collect quality visual information that can subsequently be documented and viewed to dissect events. Forensic photography can include a very wide variety of techniques and applications [1, 2], and in some special cases it has been demonstrated that extending the spectrum of wavelengths of radiation to which a camera can record information reveals previously unsuspected evidence. For example, reflected UV photography using film based methodologies [3] has been applied for forensic purposes including documenting latent bite or bruise marks, repainted surfaces, finger and shoe prints and document forgery examination [3-9].

Photography has undergone tremendous changes in the past decade because of the availability of digital cameras and computer software [10], which has also expanded the capacity for law enforcement to collect visual information [11]. However, with the exception of a few contributions [12], the use of forensic ultraviolet photography has not been widely employed to date due to the very specific technical requirements required to collect quality evidence in the UV region of the electromagnetic spectrum [13]. In particular, the characteristics of UV imaging typically result in low contrast, but high noise images that can be problematic for interpretation [3, 13]. Controlling contrast is especially important for forensic image analyses [14], and thus for the presentation of information from outside our normal visual spectrum it is vital to have calibrated reflectance scales to accurately map UV reflections to known values [13, 15]. Furthermore, since most lens material such as optical glass does not transmit UV radiation (and may even have multi-coating to exclude short wavelength radiation) it is essential to have (a) specialised optical equipment [3, 16], (b) UV rich illumination sources [3, 17], and (c) specialised filters that transmit UV but block longer wavelengths, to which digital sensors are more sensitive [13, 18]. With the requirements on the procedures of expert evidence for admission in courtrooms, it is thus essential to have standardized scientific methodology for the use of new technologies so that such information can be considered appropriately in a correct legal framework [19].

Despite the technical challenges for UV forensic imaging, which have often limited what subject matter can be easily photographically evaluated; it has recently been possible to develop techniques of using UV digital imaging for forensics in a way that is robust, reproducible, and thus applicable legal evidentiary purposes [13]. Specifically, by carefully using the physics principles involved in the production of camera responses [18], it has been possible to 'linearise' camera sensors for a FujiS3 UVIR camera, and a modified Nikon D70s camera such that the precise total amount of radiation reflected from an 'unknown sample' can be determined [13, 20]. Importantly, since the camera responses reported by most consumer level digital cameras are not linearly related to the amount of radiation reaching the sensor, these responses cannot be directly used as a measure of reflectance [21-23]. Such a non-linear relationship is introduced through image processing operations performed on the sensor's response to enhance the visual appearance of the image and to match the monitor's colour profile [18, 24, 25], prior to the production of an image which can be modified by the user. As a consequence, most of the 'raw' images available for processing present at least some degree of non-linear relationship with irradiance by the time that the user is able to obtain any information about image characteristics. Therefore, in order to use digital imaging for recording evidence in the UV region of the spectrum, it is essential to firstly establish which channels of a particular camera have sensitivity to UV radiation, and secondly make a statistically proven linear correction (Figure 1 panels a, b) of the sensor outputs so that quantitative analyses can be made [13, 21, 22, 25].

The sensitivity functions, one per channel, provides information about the specific spectral regions to which a camera responds to produce a signal. For example, consumer-level digital cameras are equipped with three colour channels: red, green and blue; each respective channel sensitive to different regions of the electromagnetic spectrum, analogous to the three photoreceptor classes present in the human visual system [26]. In the particular case of the Fuji camera, both the red and blue colour channels respond to UV-A radiation; however, these two channels are not equally sensitive (Figure 1 panel b). In the Fuji camera, the red channel presents a higher sensitivity than its blue counterpart [13]. Difference in overall sensitivity between the two channels leads to variations in the amount of noise and contrast perceivable on the images recorded by each channel as a consequence of their particular signal-to-noise ratio [18].

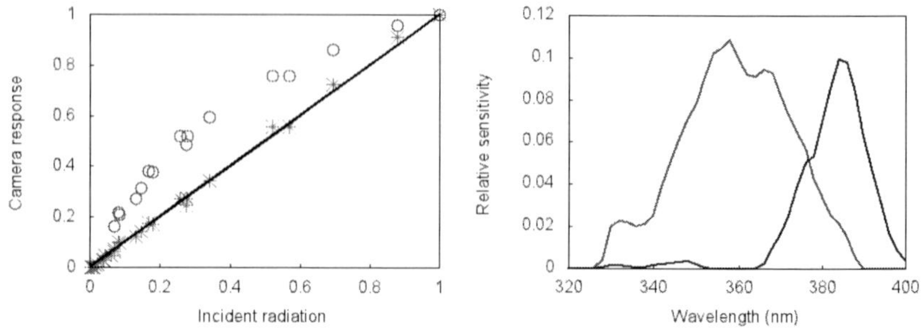

Figure 1. a) Non-linear (purple circles) and linearised camera responses (green asterisks) of the Fuji S3 UVIR camera used for our experiments to different amounts of UV-A radiation. b) Reconstructed spectral sensitivity curves for the red and blue colour channels of the same camera. Image modified from Garcia *et.al.*, 2014 [13].

Briefly, the procedure to recover linear camera responses consists on firstly recording camera responses to various levels of a known radiation source, expressed either as radiance or reflectance values [21]. Subsequently a mathematical function describing this, commonly camera-specific, relationship or its Opto-Electronic conversion function (OECF) is fitted to the data. Then the function describing the OECF is inverted for obtaining a *linearization equation*. Depending on the mathematical function selected for describing the OECF for any given camera, the linearization equation can take the form of either an algebraic operation [21], or be represented as an equation system solvable by implementing optimisation or geometrical methods [22].

The linearization equation can then be employed to recover linear camera functions, in terms of radiance or reflectance, either by solving the equation for each pixel intensity values in an image or more efficiently, by constructing a look-up table (LUT) [22]. LUTs can be easily implemented in image processing software such as Matlab (The Mathworks, USA) or ImageJ.

This new technique thus means that it will be possible to use digital UV imaging to potentially survey entire crime scenes (for example) to search for evidence that would not typically be visible to a human observer, or a regular digital camera (Figure 2). Thus by linking the benefits of digital imaging with robust methodologies that are applicable in court, the use of digital UV imaging will enable the time and cost effective collection of spectral data from a wide range of previously untested samples.

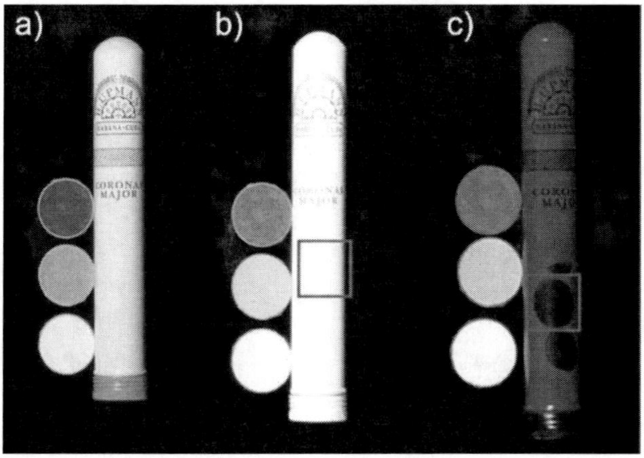

Figure 2. a) Non-linear RGB representation of a cigar canister manipulated with fingers coated with a UV absorbing substance (sunscreen) as recorded by a standard, consumer level digital camera (Canon 40D) . b) Monochrome image of a) corresponding to the image recorded by the red channel of the Canon camera. c) Linearised image corresponding to the red channel of the Fuji S3 UVIR camera after exposing the canister to UV-A radiation between 325 to 399 nm. Images adapted from Garcia et al., 2014 (13). Grey scales shown in respective images show UV-visible reflecting grey standards [15].

In the current chapter we consider an application of whether digital UV imaging could potentially be of use for imaging unknown white powders that might be present at a potential crime scene, and reliably dissecting between such white signals that otherwise look indistinguishable to the human eye or a regular digital camera. Our reason for suspecting that the UV component of the spectrum may be of value for this imaging task is that many substances tend to absorb UV, and the degree to which different 'white' samples do this may reveal important properties that might act as a useful preliminary screen for the presence of certain white powders of interest (or not) during an initial crime scene inspection.

Furthermore, in the Discussion section we discuss how this proof concept may have implications in modern forensics as a cost-effective and highly portable system that could be incorporated into modern computer tablet devices, as might be standard issue to all officers of the law that might attend a potential crime scene to conduct preliminary investigations.

MATERIALS AND METHODS

Samples

We collected a total of 19 powdered solids representing common substances found in a typical domestic situation which could potentially present as substances of interest to investigators. The entire sampling set was divided in two subsets based on the appearance of the powders to the unaided human eye: (a) a subset of 13 'white' powders, and (b) a subset of 6 'coloured' powders. Details on the selected substances are provided in Table 1.

Table 1. Details of the 19 powdered solids using during the experiment

Visual appearance	Powder Indices	Product's name	Manufacturing details
'White'	1 (a)	Automatic Dishwashing powder	Woolworths Homebrand, Bella Vista, NSW, Australia.
	2 (b)	Citric acid	Ward Mckenzie Pty. Ltd., Altona, VIC, Australia.
	3 (c)	Powder cleanser	Colgate-Palmolive Pty, Ltd., Sydney, NSW, Australia.
	4 (d)	Aspirin	Woolworths Homebrand, Bella Vista, NSW, Australia.
	5 (e)	Foot powder	Key Pharmaceuticals Pty. Ltd., Macquarie Park, NSW, Australia.
	6 (f)	Laundry powder	Colgate-Palmolive Pty, Ltd., Sydney, NSW, Australia.
	7 (g)	Iodised table salt	Woolworths Homebrand, Bella Vista, NSW, Australia.
	8 (h)	Baking powder	Ward Mckenzie Pty. Ltd., Altona, VIC, Australia.
	9 (i)	Bicarbonate of soda	Ward Mckenzie Pty. Ltd., Altona, VIC, Australia.
	10 (j)	Pure icing sugar	Sugar Australia Pty., Yarraville, VIC, Australia.
	11 (k)	Paracetamol	Woolworths Homebrand, Bella Vista, NSW, Australia.
	12 (l)	Plain flour	Woolworths Homebrand, Bella Vista, NSW, Australia.
	13 (m)	Caster sugar	Sugar Australia Pty., Yarraville, VIC, Australia.
'Coloured'	14 (n)	Mustard	Hoyts Food Ind. P/L, Moorabbin, VIC, Australia.
	15 (o)	Saffron	Hoyts Food Ind. P/L, Moorabbin, VIC, Australia.
	16 (p)	Cinnamon	Hoyts Food Ind. P/L, Moorabbin, VIC, Australia.
	17 (q)	Turmeric	Hoyts Food Ind. P/L, Moorabbin, VIC, Australia.
	18 (r)	Cayenne pepper	Hoyts Food Ind. P/L, Moorabbin, VIC, Australia.
	19 (s)	Coffee	Nestlé Australia. Ltd., Rhodes, NSW, Australia.

Camera System and Image Recording

We used a properly characterised Fuji UVIR S3 digital camera sensitive to ultraviolet (UV) (Fuji corp., Japan), visible and infrared radiations covering a spectral interval from about 325 to about 1000 nm [13]. A Micro Nikkor 105 mm quartz lens (Nikon corp., Japan) was fitted to the camera to ensure the transmission of ultraviolet radiation. The use of quartz optics in this lens allows for transmitting radiation from about 200 nm up to 1000 nm [3, 16, 27], entirely covering the UV-B and UV-A regions of the electromagnetic spectrum [28].

To prevent the recording of camera responses from visible and infrared radiation, a Baader U-filter (Company Seven, USA) was fitted in front of the quartz lens. The U-filter only allowed for the transmission of UV-A radiation by cutting off wavelengths longer than about 395 nm and up to about 1000 nm [13]. In contrast to UV-transmitting filters traditionally used for film-based UV imaging such as the Kodak Wratten 18A [3, 27], the U-filter only allows for the transmission of UV-A without leaking infrared radiation [13].

All images recorded for this study were irradiated with two flash strobes (Elinchrome, Switzerland) placed at 45° from its normal axis. Strobes were equipped with uncoated flash tubes emitting UV-A, visible and infrared radiation from 320 nm onwards. Two calibrated, grey scales uniformly reflecting about 33 % and 66 % radiation between 300 and 700 nm [15] were included in each shot to control for exposure. A white calibration standard reflecting about 99 % of incident radiation across the same spectral interval was also included in each frame as an additional exposure control.

The Fuji camera was set-up in a copy stand and the sensor plane was placed at 1,985 mm away from each of the targets. Targets were centred in the viewfinder and sharply focused before fitting the Bader U-filter since the lens is chromatically corrected visible to UV [3]. The exposure was set at 1/60 s, *f*-stop 8.0, ISO 200 after performing several exposure tests using the white calibration standard as reference.

Image Processing

Images were recorded on the native RAW format of the Fuji camera. Raw image files were then opened in the Adobe Camera Raw plug-in available in the Adobe Creative Cloud Suite (Adobe, Inc., USA), for exposure equalisation. Exposure was fine-tunned on each image using the 33 %

reflectance calibration target as a reference. The overall exposure value for each image was modified such that the mean pixel intensity value for the red channel corresponding to a 20 x 20 pixel area centred at the 33 % grey calibration target, was between 173-174 non-linear pixel intensity levels (ρ_{nl}). This exposure corresponds to a radiation of about 2.92 to 2.96 x 10^{-3} μmol, which represents 33 % of the UV-A radiation required to saturate the camera sensor before clipping occurs [29].

Exposure-calibrated images were converted into the Adobe 1998 colour space and encoded into 8-bit, uncompressed TIFF files using the same software package. An 8-bit space was used to ensure that the method could be easily utilised on all current computing platforms and software. TIFF images were then opened in Matlab release 2014b (The Matworks, USA) and linearised in terms of reflectance by using LUTs previously derived for this camera. Details on the methodology implemented for characterising the camera and constructing the LUTs are provided elsewhere [22], along with a complete mathematical description of the linearisation process.

Spectral Characterisation of the Powders

The 19 selected substances were characterised in terms of their spectral reflectance profile and their total UV-A reflectance as recorded by the UV-sensitive digital camera. Spectral reflectance curves from each substance were measured in a 300 to 710 nm spectral interval using an Ocean Optics USB 4000 spectrophotometer (Ocean Optics, USA) connected to a DH-2000 light source (Ocean Optics, USA) by means of 200 μm bifurcated optical fibre. The light source continuously emitted UV and visible radiation by simultaneously using a tungsten and deuterium lamps. Reflectance data were subsequently averaged (binned) at 5 nm intervals using custom written code for Matlab (The Mathworks, USA) release 2014b.

Experimental Design

Two types of images were recorded during the experiment. A first type consisted of five images containing the selected substances separated from one another to characterise them based on their total UV-A reflectance values. For the second type of images, four white powders were scattered on a grey fabric surface in order to evaluate the power of the digital imaging system to identify substances based their corresponding UV, linear camera responses. This test was conducted using a double blind protocol regarding substance reflectance, and position on the material.

For the first type of calibration images, the 19 substances were carefully arranged in separate piles occupying an area of about 490 mm^2 on a flat piece of box cardboard. To ensure that all substances were equally illuminated during the photographic measurements, the respective substances were arranged in cardboard pieces of about 3,240 mm^2.

For the substance identification experiment, one of the authors (AGD) mixed different amounts of four white powdered substances by arranging them in an arbitrary position. To reduce the possibility of bias during the analysis, the precise arrangement of the four substances was not revealed to the second investigator (JEG) until the image processing was finished.

All statistical analyses were performed using the R statistical package [30]. The different libraries used to perform the different tests reported in the Result section are cited as they are presented.

RESULTS

Characterisation of UV Reflective Properties of the Powdered Substances

We characterised the reflectance properties of the samples in the UV-A (300 – 399 nm) and visible region of the spectrum (400 – 700 nm) by means of spectrophotometry and UV-photography.

Most of the 'white' powders reflected some UV radiation, but differed in the amount of reflected energy at different wavelengths (Figure 3). Powders constituted by crystals or mixtures of different substances (i.e., panels a, e, g, and m) presented variations in their reflectance readings depending on the precise location of the spectrophotometer probe, with lower reflectance values associated with thinner layer of the substance. This variation was also observed in the linearised, UV-images (Figure 4 third column). On the other hand, most of the coloured powders including: saffron, cinnamon, turmeric, cayenne pepper and coffee reflected less than about 10 % of UV-A radiation.

The distribution of the total reflectance values, represented by the linear camera responses (Figure 5), also varied across the samples. For example the laundry powder (sample 6), with a mean total reflectance of about 30 %

(sample 4, Figure 5), presented a positively skewed distribution as reflectance values are truncated at 0 %.

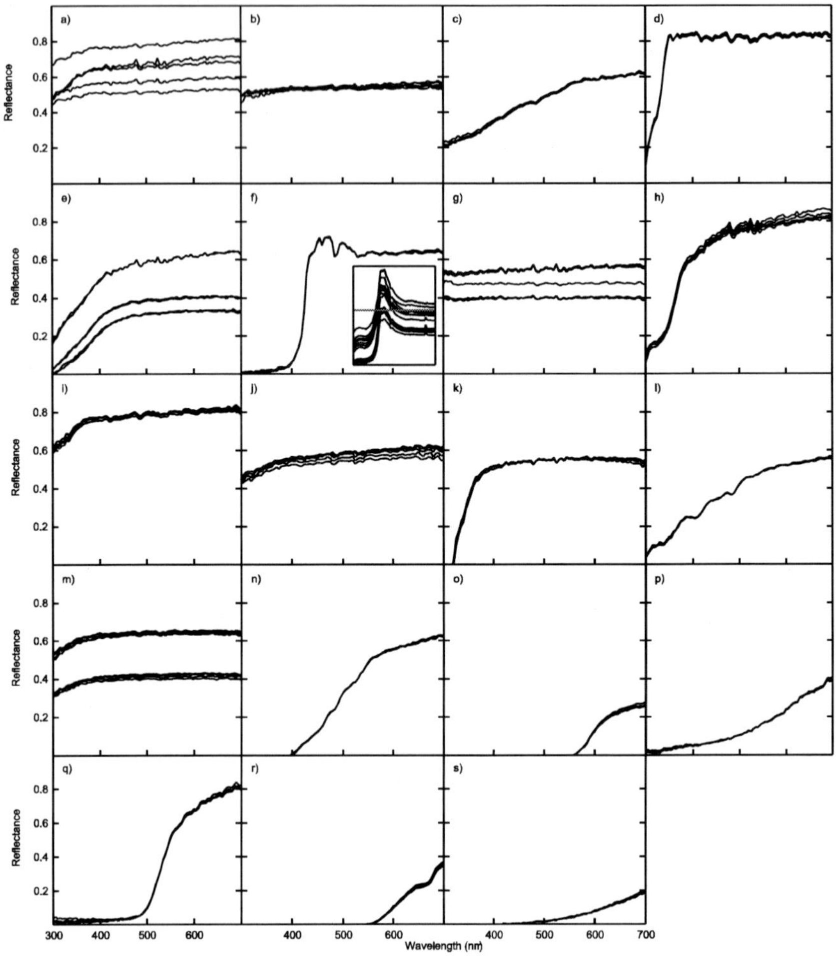

Figure 3. Spectral reflectance curves between 300 to 700 nm of the 19 powder substances using for the study. Each reflectance curve represents one point measurement obtained after placing the spectrophotometer probe at different location of the sample. Panels a) through s) correspond to substances 1 – 19 in Table 1. Panel f) depicts the spectral reflectance profile of the laundry powder, which displays a fluorescence effect when exposed to UV-radiation. Insert to panel (f) details this characteristic, as evidenced by reflectance readings above the 100 % reflectance threshold (solid red line) since this sample was highly variable due to its composite nature.

Figure 4. Photographic characterisation of the 19 solid powdered samples used during the experiment using a linearised, RGB digital camera (first column) and a linearised, UV sensitive digital camera (second column). The third column represents the same images in the second column after being pseudocoloured to facilitate the interpretation of the different reflectance values represented by the linear camera responses. Hues in panels c), f), i), l) and o) represent different reflectance values as indicated by the colour bar. Sample numbers on each chart (rows 1-5) corresponds to those identifying each substance in Table 1.

Variability is common in samples comprised of a mixture of different substances which may vary in their reflective properties as for example, by containing fluorescent components (Figure 3, panel f insert). Likewise, most of the samples presented an asymmetrical, not-normal distribution as summarised by boxplots presented in Figure 5.

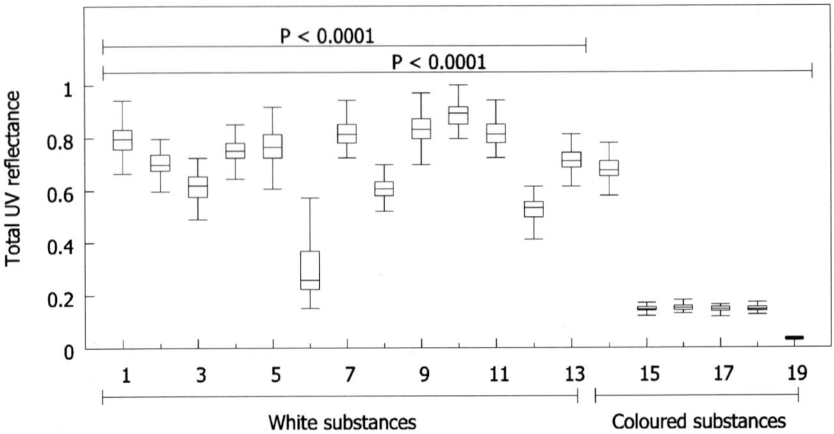

Figure 5. Boxplots summarising the distribution of the UV, linear intensity values measuring total UV reflectance for each one of the 19 powder samples using in the experiments. Significant differences in mean intensity were found across the entire sample set and across the 'white' powders. Summary was based on 100 pixel locations pseudo randomly selected from a sampling area of 2500 pixel^2 located at the centre of each powder sample.

We hypothesised that samples presenting statistical differences in their mean linear pixel values were potentially discriminable from one another. However, due to the non-normal distribution of linear camera responses for most of the samples, we implemented a non-parametrical statistical test analogous to an ANOVA [31] to test for significant differences between the different powder samples. We found statistical differences across the 19 different powders (F_t = 2110, P-values < 0.0001) including both the coloured and 'white' samples. Significant statistical differences were also found between the different 'white' powders after excluding the coloured samples (F_t = 80.4, P-value < 0.0.001).

Non-parametric *Post-hoc* pair comparisons were subsequently performed between the 19 powdered substances following the ANOVA test. Results of the pair-wise comparisons are presented as an upper triangular matrix in Table 2.

Table 2. Results of the non-parametric *post-hoc* pair comparisons between the 19 powder samples used for the experiment, based on their total UV reflectance as measured by the UV, linear pixel values. Comparisons were made using the Yuen's method with 20 % trimmed means (31). Asterisks (**) indicate significant differences at a significance level α = 0.05. NS indicates non-significant differences. Highlighted cells represent the possible combinations tested for substance identification from linear camera responses to reflected UV radiation

Substance index	1	2	3	4	5	6	7	8	9	10	11	12	13	14	15	16	17	18	19
1		**	**	NS	NS	**	NS	**	NS	**	NS	**	**	**	**	**	**	**	**
2			**	NS	NS	**	**	**	**	**	NS	**	NS	NS	**	**	**	**	**
3				**	**	**	**	NS	**	**	**	**	**	**	**	**	**	**	**
4					NS	**	**	**	**	**	NS	**	NS	**	**	**	**	**	**
5						**	NS	**	NS	**	NS	**	NS	**	**	**	**	**	**
6							**	**	**	**	**	**	**	**	**	**	**	**	**
7								**	NS	NS	NS	**	**	**	**	**	**	**	**
8									**	**	**	NS	**	NS	**	**	**	**	**
9										NS	NS	**	**	**	**	**	**	**	**
10											NS	**	**	**	**	**	**	**	**
11												**	**	**	**	**	**	**	**
12													**	**	**	**	**	**	**
13														NS	**	**	**	**	**
14															**	**	**	**.	**
15																NS	NS	**	NS
16																	NS	NS	**
17																		NS	**
18																			**
19																			

Quantitative Powder Identification from UV, Linear Camera Responses

Pair samples in Table 2 significantly different from each other are potentially discriminable based on their UV linear camera responses. To test for this hypothesis we selected samples 4 (aspirin), 6 (laundry powder), 7 (iodised salt) and, 11 (paracetamol), and arranged them in different proportions as part of a new experimental setup. This setup was photographically recorded under the same conditions as the characterisation charts (Figure 8, panel a).

To ensure an equal exposure across the characterisation and experimental images, we statistically compared the linear camera responses values (ρ) corresponding to the 33 % reflectance calibration standard. After fine tuning the exposure values, we found no significant difference in camera responses for the different images making up the first set (F = 0.412, P-value = 0.664); consequently, we pooled these values together and compared their mean camera response value ($\bar{\rho} = 173 \pm 5.42$) against the camera response value for the calibration standard in the image corresponding to the experimental setup. After calibrating the exposure of the image in the second set, we found no significant differences between the mean response values of the characterisation and experimental image (t = 0.576, P-value (two-sided) = 0.572).

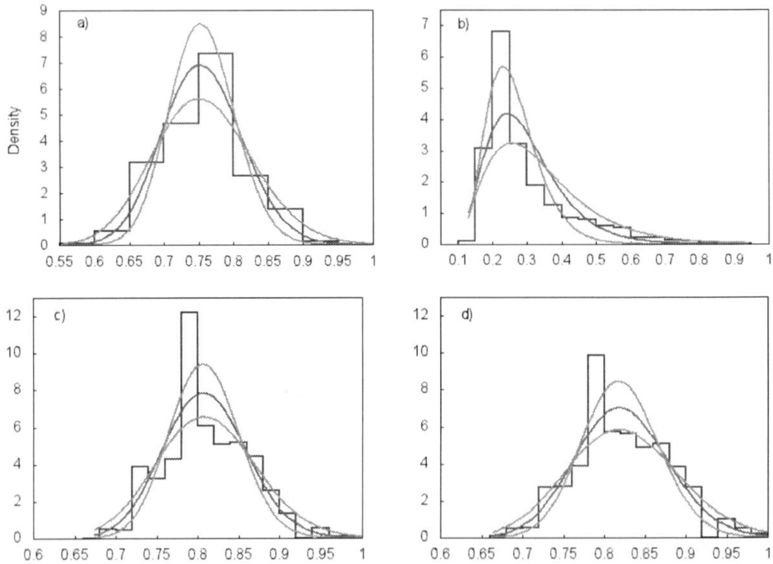

Figure 6. Histograms (magenta lines) summarising the frequency distribution of total UV response, expressed as linear, UV pixel values; for four 'white' powder samples: a) aspirin, b) laundry powder, c) Ionised salt and d) paracetamol. Solid lines represent the probability density function (PDF) fitted to each data set. Parameters describing each distribution were obtained by bootstrap using 10000 replicates. For each sample, the green solid line corresponds to the PDF reconstructed from the bootstrapped parameters; yellow and blue solid lines represent PDFs reconstructed from parameters values at the upper and lower 95 % confidence interval respectively. Details on the coefficients defining each PFD are provided in Table 3.

Image threshold is a technique used for segmenting image elements based on their intensity values.

linearised camera responses corresponding to the four substances used for the image segmentation experiment: S4 aspirin, S6 laundry powder, S7 ionised salt and S11 paracetamol. Mean and Median values are expressed in terms of linear pixel intensity values (ρ) representing total UV reflectance. Values in parentheses represent the limits of the 95 % confidence interval for each statistic and parameter. Confidence intervals were obtained from a bootstrap analysis with 10000 repetitions. [a] W-values correspond to the Anderson-Darling goodness-of-fit test for the fitted distributions (33); all obtained values were below the significance point at $\alpha = 0.05$ ($W_{\alpha\ 0.05} = 2.492$). [b] P-values corresponding to the W values were obtained from tables by Lewis 1961 [34]

	Powder sample summary statistics				Fitted probability distribution				
	Mean (ρ)	Median (ρ)	Skewness	Kurtosis	Type	Parameters	Values	W[a]	P-value[b]
S4	0.758	0.755	0.114	-0.107	Gamma	Shape	171 (112,257)	1.16	0.268
	(0.740,0.773)	(0.737,0.781).	(-0.551,0.86)	(-0.996,1.76)		Rate	226 (148,340)	(0.365,3.31)	
S6	0.301	0.254	1.50	1.93	Log-Normal	mean-log	-1.28 (-1.38,-1.17)	2.04	0.087
	(0.266,0.340)	(0.229,0.289)	0.833,2.37)	(-0.459,6.68)		sd-log	0.366 (0.290,0.437)	(0.746,4.82)	
S7	0.810	0.807	0.228	-0.497	Gamma	Shape	253 (177,363)	1.32	0.217
	(0.796,0.825)	(0.789,0.832)	(-0.256,0.761)	(-1.15,0.673)		Rate	312 (218,449)	(0.460,3.64)	
S11	0.823	0.819	0.203	-0.451	Gamma	Shape	208 (144,300)	1.26	0.232
	(0.806,0.839)	(0.797,0.841)	-(0.300,0.765)	(-1.14,0.699)		Rate	253 (175,366)	(0.430,3.55)	

The precise values representing the lower and upper threshold limits are commonly established based on image intensity histograms in such way that the most frequent intensity values for a given object are selected [32]. When pixel intensity values are distributed in a Gaussian-like (normal) manner, threshold values can be easily found based on the mean and variance values of the sample; however, this was not the case for all the samples selected (Figure 6).

In order to find the best threshold limits for identifying the different powders by image segmentation, we fitted statistical distributions other than the normal to the pixel intensity values obtained from each one of the samples. Whilst samples 4, 7 and 11 were well described by a gamma distribution, sample 6 was best described by a log-normal distribution (Table 3).

Threshold values for segmenting the four samples were established at the 0.4 and 0.6 quantiles of each fitted distribution. The use of these values ensured that only the most frequent pixel intensity values, potentially uniquely characterising each sample, were selected for image segmentation (Figure 7).

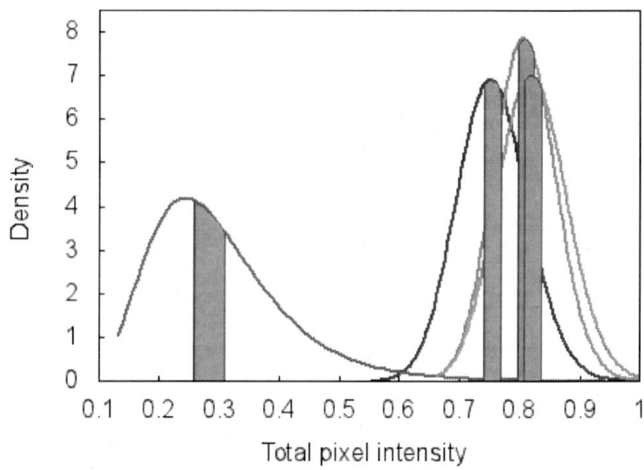

Figure 7. Probability density functions fitted to the pixel intensity values corresponding to the four samples used for the image segmentation experiment: aspirin (purple line), laundry power (green line), ionised salt (green line) and paracetamol (yellow line). Red regions indicate the 0.4 to 0.6 quantiles of the distribution of linear camera responses for each sample: aspirin (0.741-0.770), laundry powder (0.256-0.308), ionised salt (0.796, 0.822) and paracetamol (0.807-0.836). Note how the quantiles for ionised salt and paracetamol overlap; this suggests that these two samples are almost indistinguishable from one another solely based on their linearised camera responses for the channel modelled by the distributions.

After segmenting the linearised image containing powder samples 4, 6, 7 and 11 four different binary image masks were obtained; one for each respective sample. Each one of these images was subsequently pseudocoloured to visually identify the position and arrangement of the different substances in the experimental image (Figure 8).

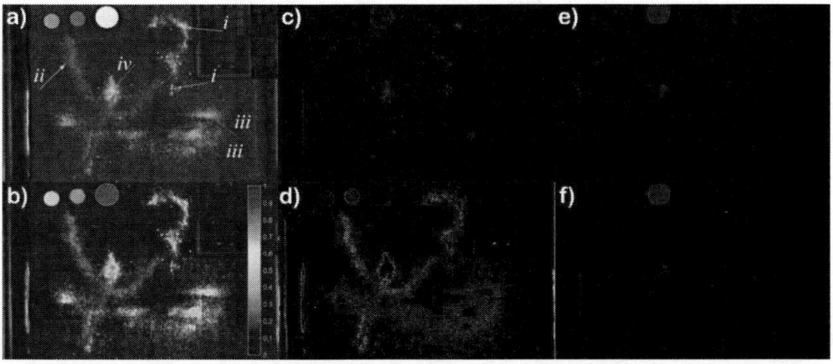

Figure 8. Experimental setup constructed for testing the power of the linarised UV recording system to identify four different substances based on reflected UV radiation. a) Linearised UV image depicting the experimental setup; arrows indicate the points of highest concentration of each substance: *i*) aspirin (cyan arrow top right corner); *ii*) laundry powder (yellow, top left corner); *iii*) ionised salt (magenta arrow lower right corner); and, *iv*) paracetamol (green arrow centre). b) Pseudocoloured representation of a) indicating total UV reflectance per pixel location within the image. Different hues indicate different amount of reflectance as indicated by the colour bar. c) Pseudocoloured, binary threshold after segmenting the image in a) using the most frequent reflectance values for aspirine. d) Pseudocoloured, binary threshold mask after segmenting the image in a) using the most frequent reflectance values for the laundry detergent. e) Pseudocoloured, binary threshold mask after segmenting the image in a) using the most frequent reflectance values for the ionised salt. f) Pseudocoloured, binary threshold mask after segmenting the image in a) using the most frequent reflectance values for paracetamol.

Our linearised camera system correctly identified between samples whose pixel intensity values corresponded to those obtained from the characterisation phase of the experiment, and those predicted from the statistical modelling of the distribution of pixel values for each sample; however, samples were not uniquely identified across the entire image. In particular the imaging system could not uniquely identify the four problem substances in image regions containing powder concentrations different to those used during the calibration stage, and for the two substances whose mean camera responses were equal in a statistical sense (Table 2).

DISCUSSION

Differential absorption and reflection of UV-A radiation by substances appearing similarly coloured in the visible region of the spectrum have been previously used for identifying substances of interest using qualitative methods [6]. Indeed, the possibility of recording reflected UV implementing photography has led to the suggestion for this technology as a potential mechanism to search remotely for evidence hard to detect or invisible such as stains, fingerprints or fainted bruises [13, 17, 35]. Nevertheless the possibility of developing a robust methodology for quantitative identification of evidence in the different UV spectral regions using imaging has been poorly studied to date due to the difficulty of properly calibrating a system for this purpose [22] and the relatively low UV-sensitivity of most current consumer-level photographic sensors. Here we used linearised camera responses as a measurement of total UV-A reflected radiation, for exploring the capacity of a UV digital imaging system for identifying powder substances, as an example, based on their differential UV-A reflection.

The capacity of imaging systems for identifying substances of interest to investigators from other non-relevant, but similarly looking substances, has great potential for developing fast and economic tools for the initial scanning of crime scenes. For example, a UV-sensitive sensor equipped with a UV transmitting lens and a UV-white light emitting diode (LED) light source [36, 37] could be attached to a tablet or mobile phone. Images recorded with the camera could be almost immediately displayed on screen using dedicated image processing software. Investigators can then use the device's touch screen for selecting objects of interest within the scene that are analysed by the software in terms of total reflectance after linearising the camera responses using look-up tables already available in the software. Total reflectance data can then be sent wirelessly to a server and compared against reference data for known substances available in police databases. With such information at their disposal, an investigator could then quickly assess the necessity of performing an exhaustive search and identification of evidence using more precise chemical methods, such as spectroscopy or chromatography [38-40], either *in situ* or at the forensic laboratory.

Methods for substance identification based on Raman spectroscopy are capable of uniquely identifying substances based on their molecular composition [39], thus becoming increasingly popular for remotely identifying either exposed or concealed hazardous, toxic and forbidden substances [38, 39]. Still, spectroscopy methods are in many cases used for examining previously identified suspicious targets, rather than for scanning for the presence or absence of potential evidence. On the other hand, image-based remote substance detection methods, as the one here presented, may potentially aid investigators to efficiently scan complex scenes facilitating the detection of suspicious targets which can subsequently be precisely identified by spectroscopy methodologies. Moreover, image-based techniques for substance identification possess several characteristics which may favour their implementation by a wider group of law enforcement groups, as for example police officers, which are commonly the first to arrive to a potential crime scene to make timely decision about the need for further investigations.

Image-based identification techniques should then be used not as a unique tool, but as part of several identification methodologies. Indeed, image-based techniques are easy to use, intuitive, practical and cost-effective making them ideal as a pre-screening or early diagnostic tool for detecting potentially useful evidence in a crime scene. Therefore, by using pre-screening imaging techniques it could be possible to maximise the use of sophisticated identification techniques which use requires a more specialised training and equipment.

Results from our substance identification paradigm indicate that there are two critical factors determining the extent to which an imaging system can uniquely identify substances based on their reflectance spectral profile: (a) the particular spectral sensitivity of the of the camera, including the number of channels available, and (b) the calibration information available to describe the reflective properties of the target sample(s).

The spectral sensitivity curves of the Fuji S3UVIR digital camera employed for our experiments indicate that the linear response of the red channel corresponds to the total energy reflected by the sample in a spectral interval of about 320 to 395 nm (Figure 1, panel b). Sensing such an interval with just single channel is not detrimental for identifying substances showing a relatively flat spectral curve in the UV region of the spectra, i.e., the ionised salt (Figure 3, panel f). In cases such as this, the availability of additional channels will not improve the system's resolution as all channels will report the same total reflectance as it is constant across the sensed spectral interval. However, for those substances presenting changes in their reflectance spectrum across the spectral interval of interest, i.e., Paracetamol (Figure 3, panel k), sensing the 320 - 400 nm spectral interval with more than one sensor provides extra information about the sample facilitating its unique identification. Moreover, such a system could potentially differentiate between two samples with similar reflectance in a single channel as for example aspiring and paracetamol (Figure 7). An imaging system sensing the UV-A spectral interval with two or more channels, a multispectral or hyperspectral system, should have a higher resolution thus increasing the possibility of uniquely identifying samples based on their UV reflective profile.

Hyperspectral and multispectral systems are currently being used increasingly for understanding the different spectral components of naturally occurring colour patterns [41]. However, their potential applications for the forensic practice remain largely unexplored. An interesting alternative could be irradiating the substances with narrow-band, quasi monochromatic radiation as for example with a polilight system. These light sources are being used in a forensic context for detecting fingerprints by induced fluorescence [42, 43] and absorption; nevertheless, their application for reflected radiation imaging has remained very limited to date because of the need to have robust methods for quantifying the data.

The second factor limiting the unique identification of substances using images is the amount of information available to characterise the samples of interest. For example, in powder substances made of crystals such as the ionised salt, the thickness of the sample affects the amount of radiation reflected and transmitted by it (Figure 3 and Figure 4). This phenomenon also occurs with semi-transparent or translucid biological samples such as thin tissue and leaves, making it difficult to adequately characterise the reflective properties of these materials [44]. Such variability is very likely to occur because as the thickness of the layer is reduced more radiation is absorbed by the background surface rather than being reflected back to the sensor. As consequence, when thin layers of this type of substances are present along with other substances reflecting lower amounts of UV radiation it is very difficult to discern between these two (Figure 8 panels b and d).

A possible solution to this problem could be including as part of the substance characterisation step a second parameter for identifying samples made of crystals or other translucid materials. In these cases, including information about the variation of UV reflectance as a function of thickness may lead to identify reflectance parameters more reliable for identifying this type of substances.

These possibilities suggest that digital UV imaging for forensics may allow for new advancements in the detection and quantification of evidence, and the rapid advancements in technology to facilitate these possibilities over the coming decade.

ACKNOWLEDGMENTS

JEG was partially supported by Colfuturo (Colombia) Crédito Beca 200818772, AGD acknowledges support by the Australian Research Council DP0878968.

REFERENCES

[1] Miller LS, Marin N, McEvoy Jr RT. Police Photography. Sixth ed. Burlington, MA: Routledge; 2010.

[2] Spring G. Forensic Photography. In: Peres MR, editor. The Focal Encyclopedia of Photography. Fourth ed. Burlington: Focal Press / Elsevier; 2007. p. 535-7.

[3] Williams AR, Williams GF. The invisible image--a tutorial on photography with invisible radiation, Part 1: Introduction and reflected ultraviolet techniques. *Journal of biological photography*. 1993;61(4):115-32.

[4] David TJ, Sobel MN. Recapturing a five-month-old bite mark by means of reflective ultraviolet photography. *Journal of Forensic Sciences*. 1994;39(6):1560-7.

[5] Krauss TC, Warlen SC. The forensic science use of reflective ultraviolet photography. *Journal of Forensic Sciences*. 1985;30(1):262-8.

[6] Richards A. Reflected Ultraviolet Imaging for Forensics Applications: Oculus Photonics; 2010 [updated March 28th; cited 2012 May 19]. 16 Pages]. Available from: http://www.company7.com/library/nikon/notes.html.

[7] Richards GB. The application of electronic video techniques to infrared and ultraviolet examinations. *Journal of Forensic Sciences*. 1977;22(1):53-60.

[8] West MH, Barsley RE, Hall JE, Hayne S, Cimrmancic M. The detection and documentation of trace wound patterns by use of an alternative light source. *Journal of Forensic Sciences*. 1992;37(6):1480-8.

[9] West MH, Billings JD, Frair J. Ultraviolet photography: Bite marks on human skin and suggested technique for the exposure and development of reflective ultraviolet photography. *Journal of Forensic Sciences*. 1987;32(5):1204-13.

[10] Salvaggio N. Basic photographic materials and processes. Third ed. Amsterdam: Focal Press/Elsevier; 2009. 462 p.

[11] Porter G. Images as Evidence. *Precedent.* 2013(119):38-42.

[12] Vogeley E, Pierce MC, Bertocci G. Experience with wood lamp illumination and digital photography in the documentation of bruises on human skin. *Archives of Pediatrics and Adolescent Medicine.* 2002;156(3):265-8.

[13] Garcia JE, Wilksch PA, Spring G, Philp P, Dyer A. Characterization of Digital Cameras for Reflected Ultraviolet Photography; Implications for Qualitative and Quantitative Image Analysis During Forensic Examination. *Journal of Forensic Sciences.* 2014;59(1):117-22.

[14] Vanderwee J, Porter G, Renshaw A, Bell M. The investigation of a relative contrast index model for fingerprint quantification. *Forensic Science International.* 2011;204(1–3):74-9.

[15] Dyer AG, Muir LL, Muntz WRA. A Calibrated Grey Scale for Forensic Ultraviolet Photography. *Journal of Forensic Sciences.* 2004;49(5):1-3.

[16] Ray S. Applied Photographic Optics. Third ed. Oxford, UK: Focal Press; 2002.

[17] Carter-Snell C, Soltys K. Forensic ultraviolet lights in clinical practice: evidence for the evidence. *Can. J. of Police & Secur Serv.* 2005;3(2):79-85.

[18] Holst GC, Lomheim TS. CMOS/CCD Sensors and Camera Systems. Bellingham, Washington, USA: SPIE Press; 2007.

[19] Merlino ML, Springer V, Kelly JS, Hammond D. Meeting the challenges of the Daubert trilogy: refining and redefining the reliability of forensic evidence. *Tulsa L Rev.* 2007;43:417.

[20] Pike T. Using digital cameras to investigate animal colouration: estimating sensor sensitivity functions. *Behavioral Ecology and Sociobiology.* 2011;65(4):849-58.

[21] Stevens M, Párraga CA, Cuthill IC, Partridge JC, Troscianko TS. Using digital photography to study animal coloration. *Biol. Jour. Lin. Soc.* 2007;90(2):211-37.

[22] Garcia JE, Dyer AG, Greentree AD, Spring G, Wilksch PA. Linearisation of RGB Camera Responses for Quantitative Image Analysis of Visible and UV Photography: A Comparison of Two Techniques. *PLoS ONE.* 2013;8(11):e79534.

[23] Cheung V, Westland S, editors. Accurate Estimation of the Non-Linearity of Input-Output Response for Color Digital Cameras. IS&T

PICS Conference; 2003; Rochester, NY, USA: The Society for Imaging Science and Technology.

[24] Allen E, Bilisi E. Digital Cameras and Scanners. In: Allen E, Triantaphillidou S, editors. The Manual of Photography. Oxford: Focal Press / Elsevier; 2011. p. 263-88.

[25] Westland S, Ripamonti C. Computational Color Science Using MATLAB. Chichester: John Wiley & Sons; 2004.

[26] Lee H-C. *Introduction to Color Imaging Science*. Cambridge: Cambridge University Press; 2005. 695 p.

[27] Eastman Kodak Company. *Ultraviolet & Fluorescence Photography*. Rochester, USA1972.

[28] Diffey BL. Sources and measurement of ultraviolet radiation. *Methods*. 2002;28(1):4-13.

[29] Garcia JE, Greentree AD, Shrestha M, Dorin A, Dyer AG. Flowers through the lens: Quantitative measurement with visible and ultraviolet digital photography. *PLoS ONE*. 2014;9(5):e96646.

[30] Core Team R. R: A Language and Environment for Statistical Computing. 3.1.3 ed. Vienna: R Foundation for Statistical Computing; 2015.

[31] Wilcox RR. Introduction to robust estimation and hypothesis testing. Third ed. St. Louis: Academic Press; 2012.

[32] Gonzalez RC, Woods RE. Digital Image Processing. Third ed: Pearson Prentice Hall; 2008.

[33] Anderson TW, Darling DA. A test of goodness of fit. *Journal of the American Statistical Association*. 1954;49(268):765-9.

[34] Lewis PA. Distribution of the Anderson-Darling statistic. *The Annals of Mathematical Statistics*. 1961:1118-24.

[35] Kaczor K, Clyde Pierce M, Makoroff K, Corey TS. Bruising and Physical Child Abuse. *Clinical Pediatric Emergency Medicine*. 2006;7(3):153-60.

[36] Nishida T, Ban T, Kobayashi N. High-color-rendering light sources consisting of a 350-nm ultraviolet light-emitting diode and three-basal-color phosphors. *Applied Physics Letters*. 2003;82(22):3817-9.

[37] Yim DK, Song HJ, Cho I-S, Kim JS, Hong KS. A novel blue-emitting NaSrPO4:Eu2+ phosphor for near UV based white light-emitting-diodes. *Materials Letters*. 2011;65(11):1666-8.

[38] Armenta S, de la Guardia M. Direct Analysis of Samples. Handbook of Green Analytical Chemistry: John Wiley & Sons, Ltd; 2012. p. 85-102.

[39] Izake EL. Forensic and homeland security applications of modern portable Raman spectroscopy. *Forensic Science International.* 2010;202(1–3):1-8.

[40] Olds WJ, Jaatinen E, Fredericks P, Cletus B, Panayiotou H, Izake EL. Spatially offset Raman spectroscopy (SORS) for the analysis and detection of packaged pharmaceuticals and concealed drugs. *Forensic Science International.* 2011;212(1–3):69-77.

[41] Chiao CC, Wickiser JK, Allen JJ, Genter B, Hanlon RT. Hyperspectral imaging of cuttlefish camouflage indicates good color match in the eyes of fish predators. *P Natl Acad Sci USA*. 2011;108(22):9148-53.

[42] Choi MJ, Smoother T, Martin AA, McDonagh AM, Maynard PJ, Lennard C, et al. Fluorescent TiO2 powders prepared using a new perylene diimide dye: Applications in latent fingermark detection. *Forensic Science International.*173(2):154-60.

[43] Seah LK, Dinish US, Phang WF, Chao ZX, Murukeshan VM. Fluorescence optimisation and lifetime studies of fingerprints treated with magnetic powders. *Forensic Science International.*152(2):249-57.

[44] Fleishman LJ, Leal M, Sheehan J. Illumination geometry, detector position and the objective determination of animal signal colours in natural light. *Animal Behaviour*. 2006;71:463-74.

Reviewed by: Associate Professor Glenn Porter, School of Photomedia, James Cook University, Queensland, Australia (glenn.porter@jcu.edu.au).

In: Forensic Science
Editor: Johanna Brewer
ISBN: 978-1-63483-086-7

Chapter 2

New Approaches in Postmortem Interval (PMI) Estimation

***Sara C. Zapico*[1,*] *and Joe Adserias Garriga*[2]**
[1]Smithsonian Institution, National Museum of Natural History, Washington DC, US
[2]University of Barcelona, Spain

Abstract

Establishing the interval between the time of death and when a body is found is one of the most complex questions to be answered by forensic scientists. Postmortem interval (PMI) estimation is absolutely crucial in forensic science for the reconstruction of perimortem events, leading to possible criminal and civil repercussions. In order to estimate accurately the time since death, it is essential to have adequate knowledge in decomposition process and the impact of the different environmental conditions in that process.

The postmortem changes useful for estimating time since death span from different processes: physical (like body cooling and hypostasis); metabolic (supravital reactions); physic-chemical (*rigor mortis*); bacterial (putrefaction); autolysis (loss of selective membrane permeability, diffusion) and insect activity.

* Corresponding author: Sara C. Zapico. E-mail: Casado-ZapicoS@si.edu; saiczapico@gmail.com.

However, as time passes, taphonomic factors accumulate their effects on the corpse, being less precise and challenging the estimation of postmortem interval. As a result, PMI is really difficult to be accurately determined, so that forensic researchers are in need of modern tools to assist in PMI estimation. Scientific efforts should be made to replace the traditional methods of estimation the time since death by those that calculate this parameter providing mean values and confidence limits, including the precision and accuracy of calculation. Continuing research is developing towards this objective.

Numerous methods have been proposed for the determination of the time since death mainly by chemical means, like the measure of volatile fatty acids in the soil solution, as well as the analysis of amino acids, neurotransmitters and decompositional by-products, leading to the development of the new field in PMI estimation called thanatochemistry.

Respect to molecular biology approaches, RNA appears as very promising target for the estimation of PMI. Its conspicuity across tissues with different degrees of exposure to diverse environmental factors and decay rates points out to more precise calibration of the PMI, by estimating and comparing its degradation rates and state across different organs of the subject. Diverse studies of post-mortem RNA degradation have shown that RNA's stability depends on the tissue and its presence also in the long-lived hard tissues of bone and teeth conducts to the possibility of extending the potential PMI estimates to longer time spans. In this line of research, it was recently demonstrated the determination of early PMI (2-6 hours after death) by analyzing the expression of two cell death proteins, FasL and PTEN, opens the door to a new approach in PMI estimation.

New growing field towards PMI estimation is the study of bacterial community changes during body decomposition. After cessation of life occurs, human immune system interrupts its function, leading to a massive release of nutrients from autolytic human cells. As a consequence of that, microbes proliferate after human death. The composition and abundance of microbial communities change as function of time and temperature and this factor could be an accurate indicator of the elapse time-since-death. This emerging line of research joins efforts with the previous new approaches for PMI estimation.

INTRODUCTION

Estimation of postmortem interval (PMI) is one of the challenges in forensic science.

This parameter is absolutely crucial for the reconstruction of perimortem events, leading to possible criminal and civil repercussions (Madea, 1993).

The study of PMI is based on the different changes that a corpse suffers after death including: physical, like body cooling and hypostasis; metabolic like supravital reactions; physicochemical like *rigor mortis*; autolysis, loss of selective membrane permeability, diffusion; bacterial processes like putrefaction (Madea, 1993; Madea et al., 1989; Madea et al., 1990; Madea et al., 1994; Madea et al., 2001) and insect activity (Amendt et al., 2004) (Figure 1). Also, other important factors to take into account for the PMI estimation are the exchange between cadaver and environment (Carter et al., 2008; Haslam and Tibbett, 2009) and trauma related to taphonomic changes (Cross and Simmons, 2010).

This methods of estimating the time since death can be divided on different categories based on the type of measurement, description of postmortem changes, influencing factors, precision of death time estimation and evidential value (Figure 2): 1. *Quantitative measurement*, mathematical description, taking into account influencing factors quantitatively, declaration of precision, proof of precision on independent material; 2. *Subjective description (grading)*, considering influencing factors, declaration of precision, proof of precision on independent material; 3. *Subjective description* of postmortem changes, influencing factors known "in principle", empiric estimations instead of statistically evaluated reference values; 4. *Subjective description*, analogous conclusions based on empiricism and assumptions; 5. *Subjective description*, velocity of progression of PMI changes entirely depending on ambient factors; due to the broad spectrum of ambient factors no sound empirical estimation is possible (Henssge and Madea, 1988; Madea, 2005; Madea and Henssge, 1988a, b).

In order to improve the methodology for PMI estimation, in the last 60 years numerous techniques were developed based on chemical changes, falling these methodologies into category 1. As a result, postmortem chemical changes provide a mathematical description of the postmortem alterations by taking into account influencing factors, such as temperature, leading to the field called "thanatochemistry" (Madea, 2005).

However, these methods, in their traditional formulations, often fail to provide appropriate mean values and confidence limits for this parameter, including overall precision and accuracy estimates to validate the method.

For that reason, current scientific efforts are seeking to improve these measurements, increasing the research developing towards this field.

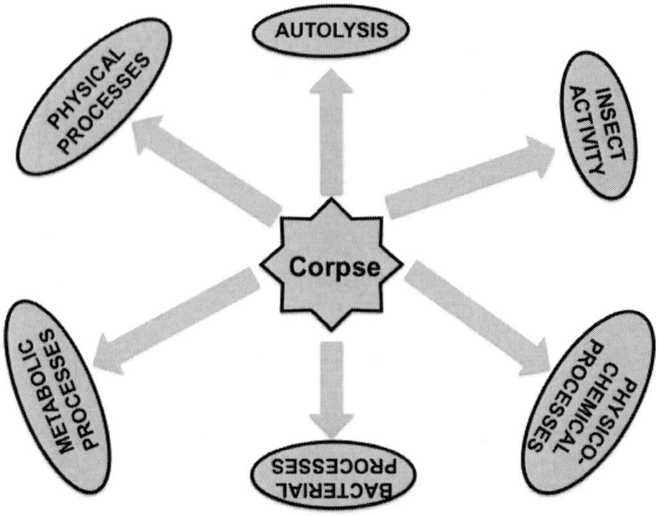

Figure 1. Type of changes that a corpse suffers after death.

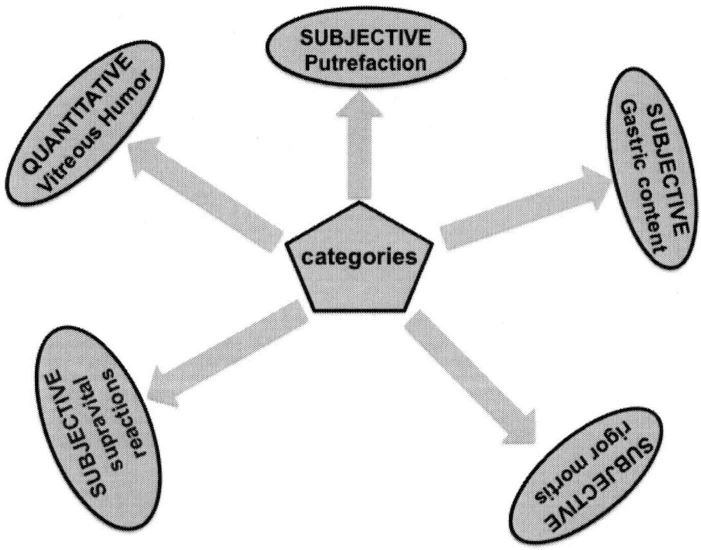

Figure 2. Types of methods for postmortem interval estimation.

Based on chemistry parameters, Vass et al. (Vass et al., 1992) used Volatile Fatty Acids (VFA) from the soil solution to determine the time since death in human cadavers.

VFA are breakdown products of both muscle and fat produced by the bacterial action on amino acids. This process is temperature dependent and because decomposition involves both aerobic and anaerobic bacteria, both processes can form these products. According to Vass et al. studies, the analysis of VFAs can yield valuable information regarding to determine time since death (TSD) and this is correlated with different decomposition stages. In fact, VFA production ceased after 1285 $\pm$ 110 Accumulated Degree Days (ADD), when the corpse is skeletonized. This research has also shown that the ions (Na^+, Cl^-, NH_4^+, K^+, Ca^{2+}, Mg^{2+}, and SO_4^{2-}) can yield valuable information regarding TSD of skeletonized human remains. The application of the formulas derived from this study to forensic cases from Tennessee showed accurate results for estimation of TSD being $\pm$ 2 days in pre-skeletonized remains and $\pm$ 2 weeks in skeletonized remains.

In this line of research, later studies of Vass et al. (Vass et al., 2002) tried to identify different biomarkers (amino acids, neurotransmitters and decompositional by-products) in various human organs (liver, kidney, heart, brain and muscle), finding different patterns depending on the organ. In this case, it is introduced the concept of Cumulative Degree Hours (CDHs), which uses a twelve hour temperature cycle to describe the decompositional process. As it was described above, different organs showed different amino acids pattern useful for PMI, however, GABA and oxalic acid have been identified as critical markers for TSD determination. In contrast, traditional decomposition markers like cadaverine and putrescine haven´t been found relevant on PMI estimation. The application of the formulas derived from this study to forensic cases from Tennessee showed accurate results for estimation of TSD.

Based on these previous studies, Vass (Vass, 2011) created two different formulas (ground human decomposition and burial decomposition) taking into account the factors that influence the rate and ultimate completeness of the decompositional process: temperature, moisture, pH and partial pressure of oxygen and knowing that soft tissue decomposition ends at 1285 $\pm$ 110 ADDs. These formulas have been found to work well in areas that comprise the mid to eastern section of the United States where humidity, soil moisture, soil type and vegetation are similar to those studied at the University of Tennessee´s Anthropology Research Facility. As a result, further research is needed towards the application of these formulas to other environments.

Apart from chemical research for the determination of TSD, many scientists have been studied the effect of PMI on DNA degradation and its use towards this objective. Dokgöz et al. (Dokgoz et al., 2001) analyzed the postmortem changes in white blood cells, finding an identifiable degeneration of neutrophils, eosinophils and monocytes 6 hours after death. In contrast, degenerative changes of lymphocytes began after 24 hours. Degenerative neutrophils were unidentifiable beyond 96 hours and eosinophils and monocytes beyond 72 hours. However, lymphocytes were still identifiable beyond 120 hours. Nuclear fragmentation started after 18 hours in neutrophils and eosinophils and after 24 hours in monocytes, while in lymphocytes occurred after 24 hours.

In this line of research, the analysis of postmortem DNA degradation by single-cell gel electrophoresis on human leukocytes (Johnson and Ferris, 2002) showed an increase in the DNA fragmentation from 2 to 22 hours. In the same study, using a porcine skeletal muscle tissue taken from 3 to 72 hours postmortem showed similar results, finding a strong correlation between DNA degradation and PMI. However, it seemed much of this degradation took place early, in the 3-24 hours postmortem period and it was not clear the death process implicated on this DNA fragmentation.

Using a computerized image analysis, Liu et al. (Liu et al., 2007) evaluated DNA degradation on spleen lymphocytes from twenty-two Sprague Dawley rats by the aid of Feulgen-van´s stained and three parameters, Integral Optical Density (IOD), Average Optical Density (AOD), Average Gray (AG). IOD and AOD tended to decrease while AG had a tendency increasing within the first 36 hours after death. However, only AG fitted and entered into the stepwise linear regression, finding a strong correlation with TSD. Though, this study had time limitations since after 36 hours autolysis made tissue smearing difficult. Previously Di Nunno et al. (Di Nunno et al., 1998) used flow cytometry to detect DNA degradation finding it specially helpful in the determination of PMI within 72 hours. After this period, DNA denaturation was massive, not allow useful measurements.

Kaiser et al. (Kaiser et al., 2008) extended these times, examined DNA degradation in human bones with PMI ranging between 1 and more than 200 years. First of all, they didn´t find a correlation between the amount of DNA and PMI. However, they demonstrated an association between presence of specific DNA products and the PMI. In this case, long fragments of β-actin-gen were analyzed, finding an inverse correlation between fragment length and PMI. Larger fragments (763 bp) could be detected within the first 8 years PMI, 507 bp fragment up to a PMI of 15 years, whereas 150 bp fragment could be amplified with the exception of one 30 and one >200 years old bone.

Beyond this studies, and center on the possibility of DNA typing, Fujita et al. (Fujita et al., 2004) found that the detection rate of each DNA types decreased with PMI, the detection rate of D1S80 typing dropped to 50% even in specimens with a short postmortem interval (>1 month).

In the way of using biomolecules for determination of TSD, some researchers analyzed the potential use of proteins. Sabucedo et al. (Sabucedo and Furton, 2003), using Western blot techniques, found a pseudo-linear relationship between percent cardiac Troponin I degraded and the log of the time since death in human cadavers, being useful in the determination of the early PMI (0-5 days).

However, other scientists are more focused on the brain (Finehout et al., 2006), applying 2-dimensional gel electrophoresis, Finehout et al. found fourteen protein from cerebrospinal fluid (CSF) that showed an increase in production that correlated with PMI and several of these proteins have been shown to play a role in protecting cells from oxidative stress. Related to this study, Chandana et al. (Chandana et al., 2009) demonstrated an increase on protein oxidation/nitration events, glial fibrillary acidic protein (GFAP) and neurofilament (NF) expression with increasing PMI on substantia nigra (SN) from human brains. Though, these previous studies on human brains were developed for the potential applications to clinical research, not by means of TSD determination.

Kikuchi et al. (Kikuchi et al., 2010) used high mobility group box-1 (HMGB1) protein towards this objective on blood samples Wistar rats by ELISA technique, finding a time-dependent increase of this protein. However, this depends on the stored temperature. At 4°C this increase was constant up to seven days; at 14°C there was a peak at day 3, then decrease at day 4 and then plateau phase. At 24°C, the results were similar to 14°C, but the peak was at day 2, decrease at day 3 and then plateau phase.

Following the line of using biomolecules to estimate the PMI, RNA appears as a very promising target. Its conspicuity across tissues with different degrees of exposure to diverse environmental factors and decay rates opens the door for more precise calibration of the TSD, by estimating and comparing its degradation rates and state across different organs of the subject. After death RNA is degraded by ribonucleases already present in the cell and/or originating from bacteria or other environmental contamination (Bauer, 2007). Based on this premise, several authors analyzed the degradation profile of RNA respect to PMI.

Inoue et al. (Inoue et al., 2002) studied the degradation of RNA of brain, lung, liver and heart from twenty-four Wistar rats after 0-7 days postmortem. First, they found that total RNA bands were more detectable in the brain, then lung and heart and less detectable in the liver.

Respect to the detection of housekeeping genes, like glyceraldehyde-3-phosphate dehydrogenase (GAPDH), using Northern blot analysis, they found a similar pattern than total RNA, on brain the expression of this protein can be detectable until 7 days postmortem. Using Quantitative-PCR (QPCR), the results were similar than the ones described above. The stability of GAPDH mRNA decreased in the following order: brain, heart, lung and liver. Apart from that, the degradation rates of GAPDH and other two housekeeping genes, β-actin and hypoxanthine guanine phoshoribosyltransferase (HPRT) in the brains were similar, suggesting that the mRNA of constitutively expressed genes degrade at a similar rate. Besides, the authors found a strong positive correlation between GAPDH values and PMI, giving the formulas for TSD determination. These results are in agreement with previous studies (Johnson et al., 1986) reported that there is not degradation of mRNA in the brain tissue from autopsy materials until 36 hours postmortem.

Later studies developed on human blood samples from living people stored from 1-4 days and postmortem blood and brain tissue samples with PMI ranges between 7 and 145 hours, did not show significant time-dependent changes on RNA amounts. In contrast, multiplex PCR of Fatty acid synthase-messenger RNA (FASN-mRNA), a housekeeping gene showed a time-dependent decrease of the fraction of intact mRNA in whole blood samples from living individuals, found a strong correlation with the storage time. Postmortem blood samples as well as brain tissue samples showed approximately similar results, a decreased on intact fraction up to a PMI of 120 hours, and similar strong correlation.

Van Doorn et al. (van Doorn et al., 2011) used bone and bone marrow to study the stability of postmortem RNA, taking samples from 0 to 31 days after death. They found differences between two tissues, through bioanalyzer analysis, rRNA peaks from bone samples could be observed up to 4 days. In contrast, rRNA from leukocyte fraction of bone marrow was visible up to 21 days postmortem. Using QPCR for three housekeeping genes showed that relative quantification were higher in the bone marrow fractions, particularly on leukocyte fractions. This study pointed out the potential use of bone marrow as source for RNA in forensic studies.

In spite of the previous work done on hard tissue, the majority of RNA studies were developed in soft tissues. Li et al. (Li et al., 2014) used Sprague-Dawley rat heart tissues collected until 168 hours after death to analyze the expression of 18S-rRNA and microRNA. They found that the level of 18S-rRNA gradually increased in the early stage and then increased in correlation with PMI, finding a parabola equation with a strong positive correlation.

They also studied three different microRNA, being the most stable miR-1 and finding a strong positive correlation between 18S-rRNA/miR-1 and PMI. This group followed these studies (Lv et al., 2014) on Sprague-Dawley rat spleen tissues collected until 144 hours after death keeping at room temperature and also until 312 hours after death keeping at 4ºC. Apart from analyzed the expression of 18S-rRNA and microRNAs (miR-125b and miR-143), they studied other endogenous molecular biological markers like GAPDH, β–actin (ACTB), on two different regions (5´-end and 3´-end), creating two pairs of primers for each marker (GAPDH1 and GAPDH2; ACTB1 and ACTB2). Besides, they used U6 small nuclear RNA (U6) as RNA marker. Using Q-PCR, on housekeeping genes, they found GAPDH1 and ACTB1 fluctuated slightly while GAPDH2 and ACTB2 decreased rapidly. microRNAs were used as control endogenous markers since they showed more stability. Based on that, they found strong positive correlations between GAPDH2, ACTB2, U6, 18S-rRNA and the TSD, although the mathematical models were different depends on the temperature.

Following the line of using gene expression to determine the PMI, Sampaio-Silva et al. (Sampaio-Silva et al., 2013) studied the expression of eleven genes on eight organs (skin, heart, spleen, femoral quadriceps, liver, pancreas, stomach and lungs) taken from Balb/c mice at 4 or 20 hours postmortem. First of all, they found that RNA from heart, spleen and lung showed the highest stability followed by femoral quadriceps, liver and stomach, being less stable pancreas and skin. Respect to QPCR analysis of genes, only on femoral quadriceps was found four genes significantly correlated with the PMI (GAPDH, ACTB, Cyclophilin A (Ppia) and Signal recognition particle 72 (Srp72)) and two genes in liver (Serum albumin (Alb) and cytochrome P450 2E1 (Cyp2E1)). Based on the results obtained from femoral quadriceps, they created a mathematical model to estimate the PMI with a confidence interval of $\pm$ 51 minutes at 95%.

Studies described above used housekeeping genes to determine the PMI. Recent research from our group (C. Zapico et al., 2014) took into account one of the earliest process in decomposition, the autolysis and the possibility of using the expression of cell death signaling proteins as markers of early postmortem interval. Taking samples of gastrocnemious muscle from four Wistar rats from 0 to 8 hours after death each 2 hours, we studied the mRNA expression of Fas Ligand (FasL), a death receptor from extrinsic apoptotic pathway, and Phosphatase and Tensin homologue deleted on chromosome 10 (PTEN), a regulator protein, by QPCR, finding an increase in the mRNA levels of both proteins up to 6 hours after death, and then decline.

Through a regression analysis of these first 6 hours, a strong positive correlation was found between mRNA expression and TSD. The decrease of the expression levels of both proteins could have two explanations: the degradation of RNA postmortem or because of death signaling itself, which produced the increase of free radicals, which in turn induce a deactivation of PTEN and as a result cannot activate FasL. In spite of this hypothesis, this line of study requires more research including additional cell death markers and extending the time period.

Apart from aforementioned studies developed by forensic means, some studies were focus on RNA value as a potential tool for clinical diagnosis, finding the same results as previously described, brain showed much greater stability up to 96 hours postmortem (Cummings et al., 2001; Iwamoto et al., 2006; Johnson et al., 1986; Johnston et al., 1997; Preece and Cairns, 2003; Schramm et al., 1999; Trotter et al., 2002). The same happens with myocardial tissue, finding that the expression of some cardiac proteins changed with the PMI (Gonzalez-Herrera et al., 2013), which could be important to understand the underlying mechanisms involved in myocardium ischemia and its repair. In the line of using RNA for research, Gopee and Howard (Gopee and Howard, 2007) demonstrated the RNA stability on postmortem skin from SKH-1 female hairless mice up to 60 minutes after death. In the same way, and related to reproductive research, Fitzpatrick et al. (Fitzpatrick et al., 2002) revealed that total RNA yields from bovine ovarian, oviduct and uterine tissue samples were stable up to 96 hours after death, finding the same results when they analyzed the expression of housekeeping genes: GAPDH, β-actin and Transforming Growth Factor-β (TGF-β) by QPCR. In contrast 28S-rRNA remained intact up to 24 hours after death, as well as the Northern blot analysis of GAPDH revealed evidence of mRNA degradation from 48 hours postmortem. These results indicated that is possible to use RNA from reproductive tissue up to 24 hours after death.

Although these previous researches related to RNA were developed by forensic or clinical means, both contributed to expand the studies related to PMI estimation based on this biomolecule, increasing the knowledge in this field.

In a macroscopic level, Payne proposed in 1965 six stages of decomposition: fresh, bloated, active decay, advanced decay, dry, remains (Payne, 1965). This is still valid nowadays and useful as macroscopic descriptor. In other cases, the corpse can experience a process of conservation (like mummification or saponification) that will not allow the body breakdown, resulting on the preservation of the corpse temporally or over time (Figure 3).

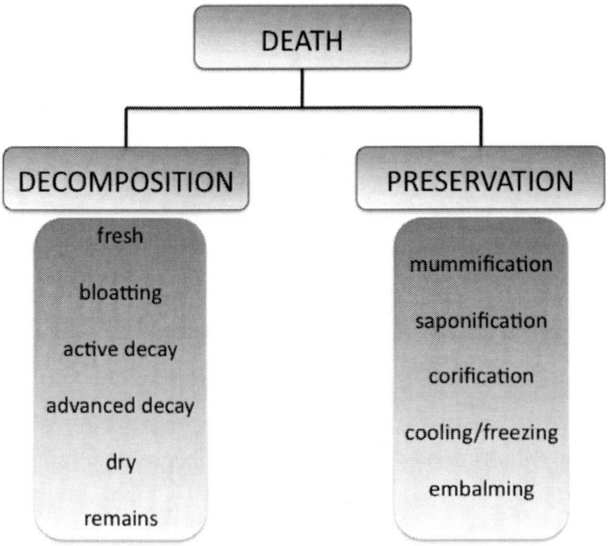

Figure 3. Cadaver decomposition stages and possible situations of body preservation.

Decomposition varies based on macroscopic factors like environment, if the body is clothed or naked, death circumstances, the place where the body is found as well as the weather. It may also vary within the same body according to "microenvironments" developed around the corpse (Schmitt et al., 2006). And even in cases where the cadaver breakdown progresses in a predictable succession of the different stages of decomposition, the duration of these stages may vary dramatically from one cadaver to another, depending specially on the environmental conditions. Because of this great variation of decomposition process, those macroscopic descriptors cannot estimate the PMI accurately, with the precision required in forensic cases. For that reason, it is important to take into account other extrinsic factors on the corpse breakdown such as insect and scavenger activity.

Insects can access to the cadaver within seconds after cessation of life (Mann et al., 1990), or even in the agony. Blowflies and flesh flies tend to dominate the early stages of corpse decomposition in an attempt to find a suitable resource for the development of their offspring. The activity of these insects can have a significant effect on cadaver decomposition (Bird and

Castner, 2010). Maggot activity can represent the primary driving force behind the removal of soft tissues (Tibbett and Carter, 2008). The excess of gas in the bloating stage can compromise the integrity of the skin, driving the cadaver into the advanced decay stage of decomposition, until skeletal remains.

Insect succession has been largely used to estimate the PMI. The most important implication for PMI estimation is that carrion insect species differ in terms of growth rate, arrival time and position within the order of succession (Bird and Castner, 2010).

There are certain concerns for the application of forensic entomology as PMI indicator, like the inaccuracy of the elapse of time between death and egg deposition (Tomberlin et al., 2011), the lack of insects during particular weather or season (Archer and Elgar, 2003), or region specific blowfly larval growth curves and insect communities (Gallagher et al., 2010). All of these factors pointed out that there are cases and situations where forensic entomology cannot be applied to estimate the PMI.

During active decay of a corpse, the decomposition island is formed around the body through an intense pulse of water, carbon and nutrients. The dynamics of the decomposition island is poorly known, although it changes over time (Towne, 2000).

These changes can be defined by a succession of insects (Kocárek, 2003), plant (Towne, 2000), and fungal (Carter and Tibbett, 2003) communities as well as variation on the concentration of chemical compounds (Vass et al., 1992). These phenomena are likely related to the physicochemical changes along the different stages of cadaver decomposition (Tibbett and Carter, 2008). All of those changes can be reflected in bacterial communities succession on the decomposition process.

Microbial activity has been known to remain active months after the end of the putrefaction stage due to the increase of carbon level, pH and nutrient concentration in soil (Vass et al., 1992). Although the bacterial role is still unclear, there is not standard method developed yet to use microbes as PMI estimator, different research studies are focused on that topic.

At least ten times more bacteria than human cells in the body inhabit the human body (NIH HMP Working Group, 2009). This great representation of bacterial cells was reflected in the concept of human microbiome, first suggested by Lederberg and McCray in 2001. This means the ecological community of commensal, symbiotic and pathogenic microorganisms that share our body space (Lederberg and McCray, 2001).

There are four major sites of microbial colonization in the human body: mouth, gut, vagina and skin (NIH HMP Working Group, 2009) (Figure 4).

Microbial colonization of the host begins at birth. The most significant changes in human microbiome take place during the three first years of life (Erturk-Hasdemir and Kasper, 2013), but microbiological changes will continue throughout life, and also after death.

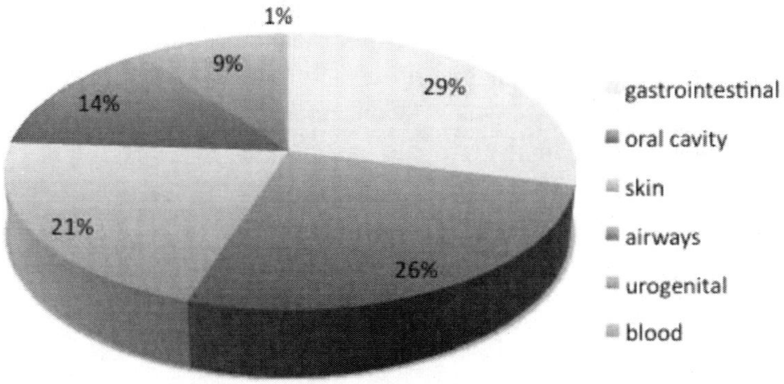

Modified from NIH HMP Working Group, 2009.

Figure 4. Bacterial distribution in different sites of the human body.

Commensal bacteria play an important role in maintaining immune homeostasis and promoting the host´s health (Erturk-Hasdemir and Kasper, 2013).

Composition of the microbiota may be altered by various external parameters including abuse of antibiotics or changes in diet, being more prone to colonization by specific pathogens. Besides, commensal bacteria can become pathogenic in case of their growth rate raises and if they outcompete other members of the microbiome (Ribet and Cossart, 2015).

In the living human body, there are different lines of defense against pathogens like epithelial barrier or the immune system. When the cessation of life occurs, those barriers do not play their functions and the corpse becomes a large nutrient resource that can support a complex and phylogenetically diverse community of organisms (Gunn and Pitt, 2012; Mondor et al., 2012). While microbiological communities recycle different nutrients in their role as decomposers, the different stages of decomposition (fresh, bloating, active decay, advanced decay, dry and remains) reveal significantly the effects of bacteria metabolism in the corpse.

Traditional microbiology culture methods have identified particular bacteria present in human decomposition process (Vass 2001, Janaway et al., 2009). Though very useful, traditional methods cannot yield an accurate account of the species diversity in a particular ecosystem, because only cultivable bacteria can be detected. It is estimated that only about 1% of bacteria can be cultured (Hyde et al., 2013; Amann et al 1995).

To address this concern, a molecular approach has been developed in the last years studies focused on postmortem microbial diversity.

Microbial ecology of the body decomposition can nowadays be researched in detail using sequencing technologies that enable the entire communities to be characterized across the timeline of decomposition (Metcalf et al., 2013). These data will not only allow us to understand the underlying microbial ecology of corpse decomposition, but also offers the possibility to use microbes as evidence (Pechal et al., 2013). The changes in microbial communities after death can be used as an estimator of PMI.

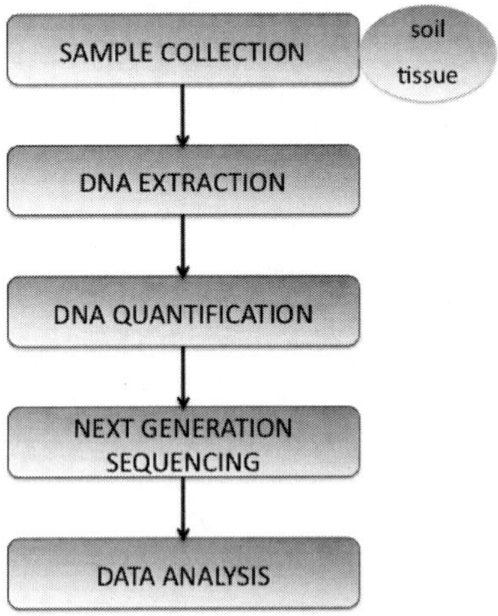

Figure 5. Present methodology used to study the microbial communities changes in decomposition process.

The methodology consists basically in sample collection, which can be taken from the soil and swapping, scratching or dissecting the body tissues to be studied. DNA is then extracted from the samples, and next generation sequencing is applied on the DNA. After that, data analysis must be carried out to interpret the meaning of the bacterial community changes in the decomposition process (Figure 5). In that sense, facultative anaerobic bacteria, such as *Lactobacillus*, predominate in organ tissues and blood samples (and therefore, they are indicative) of cadavers with short PMI, while obligate anaerobic bacteria such as *Clostridium*, predominate in corpses with longer

PMI (Can et al., 2014). Different studies have registered a shift in postmortem microbial communities, being dominated firstly by aerobic bacteria such as *Staphylococcus*, to being dominated later by anaerobic bacteria such as *Clostridium* and *Bacteroides* (Janaway et al., 2009; Carter et al., 2008; Melvin et al, 1984).

Table 1. Summary of the new approaches for PMI estimation

ANALYTE	TISSUE TYPE	TECHNIQUE AND TARGET
VFA	Soil	GC-FID
Anions	Soil	HPLC
AA, NT, Decompositional by-products	Human liver, kidney, heart, brain, muscle	GC/MS
Postmortem changes	Blood-white cells	Microscopy
DNA degradation	Human leukocytes	Single cell electrophoresis
	Porcine skeletal muscle	Single cell electrophoresis
	Spleen lymphocytes rats	Computered image analysis
	Spleen lymphocytes	Flow cytometry
	Human bones	PCR
	Human bones, nails and blood	PCR
Proteins	Human cardiac tissue	Western Blot: Troponin
	Human Cerebrospinal Fluid	2-Dimmensional Electrophoresis
	Human Substancia Nigra	Different Techniques
	Rat Blood	ELISA: HMGB
RNA Degradation	Rat: brain, heart, lung, liver	QPCR: GAPDH, beta-actin, HPRT
	Human blood, brain tissue	Multiplex: FASN-mRNA
	Rabbit bone and bone marrow	Bioanalyzer, QPCR
	Rat heart	QPCR: 18S-rRNA
	Rat spleen	QPCR: 18S-rRNA, U6, GAPDH, beta-actin
	Mice femoral quadriceps	QPCR: GAPDH, beta-actin, Ppia, Srp72
	Rat gastrocnemius	QPCR: PTEN, FasL
Insect Activity	Corpse	Insect sucession
Microbial Communities	Soil/Whole corpse	Next Generation Sequencing

Using changes in the microbiome structure to monitor the progression of decomposition process may circumvent limitations of the entomology application, as microbes are ubiquitous in the environment, located on humans before death (Metcalf et al., 2013).

As a result, concerns on the application of entomology due to the uncertain elapse of time between death and ovoposition, the absence of insects in certain weather conditions or the limited access to the cadaver may be addressed by the use of microbes as PMI estimator.

There are several new approaches towards the accurate determination of PMI. However, the application of one methodology or another depends on the forensic context and the human remains available (Table 1).

References

Amann, R. I., Ludwig, W., Schleifer, K. H., 1995. Phylogenetic identification and in situ detection of individual microbial cells without cultivation. *Microbiology Reviews* 59, 143-169.

Amendt, J., Krettek, R., Zehner, R., 2004. Forensic entomology. *Die Naturwissenschaften* 91, 51-65.

Archer, M. S., Elgar, M. A., 2003. Yearly activity patterns in southern Victoria (Australia) of seasonally active carrion insects. *Forensic science international* 132, 173-176.

Bauer, M., 2007. RNA in forensic science. Forensic science international. *Genetics* 1, 69-74.

Bird, J., Castner, J., 2010. Forensic entomology. *The utility of arthropods in legal investigations.* 2nd edition. CRC Press. Boca Raton.

C. Zapico, S, Menendez, S. T., Nunez, P., 2014. Cell death proteins as markers of early postmortem interval. *Cellular and molecular life sciences: CMLS* 71, 2957-2962.

Can, I., Javan G., Pozhitkov, A., Noble, P., 2014. Distinctive thanatomicrobiome signatures found in the blood and internal organs of humans. *Journal of Microbiological Methods* 106. 1-7.

Carter, D. O., Tibbett, M., 2003. Taphonomic mycota: fungi with forensic potential. *Journal of forensic sciences* 48, 168-171.

Carter, D. O., Yellowlees, D., Tibbett, M., 2008. Using ninhydrin to detect gravesoil. *Journal of forensic sciences* 53, 397-400.

Chandana, R., Mythri, R. B., Mahadevan, A., Shankar, S. K., Srinivas Bharath, M. M., 2009. Biochemical analysis of protein stability in human brain collected at different post-mortem intervals. *The Indian journal of medical research* 129, 189-199.

Cross, P., Simmons, T., 2010. The influence of penetrative trauma on the rate of decomposition. *Journal of forensic sciences* 55, 295-301.

Cummings, T. J., Strum, J. C., Yoon, L. W., Szymanski, M. H., Hulette, C. M., 2001. Recovery and expression of messenger RNA from postmortem human brain tissue. *Modern pathology: an official journal of the United States and Canadian Academy of Pathology*, Inc. 14, 1157-1161.

Di Nunno, N. R., Costantinides, F., Bernasconi, P., Bottin, C., Melato, M., 1998. Is flow cytometric evaluation of DNA degradation a reliable method to investigate the early postmortem period? *The American journal of forensic medicine and pathology* 19, 50-53.

Dokgoz, H., Arican, N., Elmas, I., Fincanci, S. K., 2001. Comparison of morphological changes in white blood cells after death and in vitro storage of blood for the estimation of postmortem interval. *Forensic science international* 124, 25-31.

Erturk-Hasdemir, E., Kasper, D., 2013. Resident commensals shaping immunity. *Current Opinions Immunology* 25, 450-455.

Finehout, E. J., Franck, Z., Relkin, N., Lee, K. H., 2006. Proteomic analysis of cerebrospinal fluid changes related to postmortem interval. *Clinical chemistry* 52, 1906-1913.

Fitzpatrick, R., Casey, O. M., Morris, D., Smith, T., Powell, R., Sreenan, J. M., 2002. Postmortem stability of RNA isolated from bovine reproductive tissues. *Biochimica et biophysica acta* 1574, 10-14.

Fujita, Y., Kubo, S., Tokunaga, I., Kitamura, O., Gotohda, T., Ishigami, A., 2004. Influence of post-mortem changes on DNA typing (D1S80, TH01, HLA DQA 1, and PM typing system): case studies for personal identification. *Leg. Med.* (Tokyo) 6, 143-150.

Gallagher, M. B., Sandhu, S., Kimsey, R., 2010. Variation in developmental time for geographically distinct populations of the common green bottle fly, Lucilia sericata (Meigen). *Journal of forensic sciences* 55, 438-442.

Gonzalez-Herrera, L., Valenzuela, A., Marchal, J. A., Lorente, J. A., Villanueva, E., 2013. Studies on RNA integrity and gene expression in human myocardial tissue, pericardial fluid and blood, and its postmortem stability. *Forensic science international* 232, 218-228.

Gopee, N. V., Howard, P. C., 2007. A time course study demonstrating RNA stability in postmortem skin. *Experimental and molecular pathology* 83, 4-10.

Gunn, A., Pitt, S., 2012. Microbes as forensic indicators. *Tropical Biomedicine* 29, 311-30.

Haslam, T. C., Tibbett, M., 2009. Soils of contrasting pH affect the decomposition of buried mammalian (Ovis aries) skeletal muscle tissue. *Journal of forensic sciences* 54, 900-904.

Henssge, C., Madea, B., 1988. Determination of the time since death. I. Body heat loss and classical signs of death. An integrated approach. *Acta medicinae legalis et socialis* 38, 71-89.

Hyde, E., Haarmann, D., Lynne, A., Bucheli, S., Petrosino, J., 2013. The Living Dead: Bacterial Community Structure of a Cadaver at the Onset and End of the Bloat Stage of Decomposition. *PLoS ONE* 8, e77733.

Inoue, H., Kimura, A., Tuji, T., 2002. Degradation profile of mRNA in a dead rat body: basic semi-quantification study. *Forensic science international* 130, 127-132.

Iwamoto, K., Bundo, M., Ueda, J., Kato, T., 2006. Expression of ribosomal subunit genes increased coordinately with postmortem interval in human brain. *Molecular psychiatry* 11, 1067-1069.

Janaway, R., Percival, S., Wilson, A., 2009. Decomposition of Human Remains. In: Percival, S., *Microbiology and Aging: Humana Press*. pp. 313-334.

Johnson, L. A., Ferris, J. A., 2002. Analysis of postmortem DNA degradation by single-cell gel electrophoresis. *Forensic science international* 126, 43-47.

Johnson, S. A., Morgan, D. G., Finch, C. E., 1986. Extensive postmortem stability of RNA from rat and human brain. *Journal of neuroscience research* 16, 267-280.

Johnston, N. L., Cervenak, J., Shore, A. D., Torrey, E. F., Yolken, R. H., 1997. Multivariate analysis of RNA levels from postmortem human brains as measured by three different methods of RT-PCR. Stanley Neuropathology Consortium. *Journal of neuroscience methods* 77, 83-92.

Kaiser, C., Bachmeier, B., Conrad, C., Nerlich, A., Bratzke, H., Eisenmenger, W., Peschel, O., 2008. Molecular study of time dependent changes in DNA stability in soil buried skeletal residues. *Forensic science international* 177, 32-36.

Kikuchi, K., Kawahara, K. I., Biswas, K. K., Ito, T., Tancharoen, S., Shiomi, N., Koda, Y., Matsuda, F., Morimoto, Y., Oyama, Y., Takenouchi, K., Miura, N., Arimura, N., Nawa, Y., Arimura, S., Jie, M. X., Shrestha, B., Iwata, M., Mera, K., Sameshima, H., Ohno, Y., Maenosono, R., Tajima, Y., Uchikado, H., Kuramoto, T., Nakayama, K., Shigemori, M., Yoshida, Y., Hashiguchi, T., Maruyama, I., 2010. HMGB1: A new marker for estimation of the postmortem interval. *Experimental and therapeutic medicine* 1, 109-111.

Kocárek, P. 2003. Decomposition and Coleoptera succession on exposed carrion of small mammal in Opava, Czech Republic. *European Journal Soil Biology* 39, 31-45.

Lederberg, J., McCray, A. T., 2001. 'Ome Sweet 'Omics—a genealogical treasury of words. *Scientist* 15: 8.

Li, W. C., Ma, K. J., Lv, Y. H., Zhang, P., Pan, H., Zhang, H., Wang, H. J., Ma, D., Chen, L., 2014. Postmortem interval determination using 18S-rRNA and microRNA. *Science and justice: journal of the Forensic Science Society* 54, 307-310.

Liu, L., Shu, X., Ren, L., Zhou, H., Li, Y., Liu, W., Zhu, C., Liu, L., 2007. Determination of the early time of death by computerized image analysis of DNA degradation: which is the best quantitative indicator of DNA degradation? *Journal of Huazhong University of Science and Technology.* Medical sciences = Hua zhong ke ji da xue xue bao. Yi xue Ying De wen ban = Huazhong keji daxue xuebao. Yixue Yingdewen ban 27, 362-366.

Lv, Y. H., Ma, K. J., Zhang, H., He, M., Zhang, P., Shen, Y. W., Jiang, N., Ma, D., Chen, L., 2014. A time course study demonstrating mRNA, microRNA, 18S rRNA, and U6 snRNA changes to estimate PMI in deceased rat's spleen. *Journal of forensic sciences* 59, 1286-1294.

Madea, B., 1993. Time since death extrapolated from vitreous potassium. *Forensic science international* 59, 80-82.

Madea, B., 2005. Is there recent progress in the estimation of the postmortem interval by means of thanatochemistry? *Forensic science international* 151, 139-149.

Madea, B., Henssge, C., 1988a. Determination of the time since death. II. Electrical excitability of skeletal muscles. Current state and recent developments. *Acta medicinae legalis et socialis* 38, 91-108.

Madea, B., Henssge, C., 1988b. Determination of the time since death. III. Potassium in vitreous humour. Rise of precision by use of an "inner standard". *Acta medicinae legalis et socialis* 38, 109-114.

Madea, B., Henssge, C., Honig, W., Gerbracht, A., 1989. References for determining the time of death by potassium in vitreous humor. *Forensic science international* 40, 231-243.

Madea, B., Herrmann, N., Henbge, C., 1990. Precision of estimating the time since death by vitreous potassium--comparison of two different equations. *Forensic science international* 46, 277-284.

Madea, B., Kaferstein, H., Hermann, N., Sticht, G., 1994. Hypoxanthine in vitreous humor and cerebrospinal fluid--a marker of postmortem interval and prolonged (vital) hypoxia? Remarks also on hypoxanthine in SIDS. *Forensic science international* 65, 19-31.

Madea, B., Kreuser, C., Banaschak, S., 2001. Postmortem biochemical examination of synovial fluid--a preliminary study. *Forensic science international* 118, 29-35.

Mann, R. W., Bass, W. M., Meadows, L., 1990. Time since death and decomposition of the human body: variables and observations in case and experimental field studies. *Journal of forensic sciences* 35, 103-111.

Melvin, J. R. Jr., Cronholm, L. S., Simson, L. R. Jr., Isaacs, A. M., 1984. Bacterial transmigration as an indicator of time of death. *Journal of Forensic Sciences* 29, 412-417.

Metcalf, J. L., Wegener Parfrey, L., Gonzalez, A., Lauber, C. L., Knights, D., Ackermann, G., Humphrey, G. C., Gebert, M. J., Van Treuren, W., Berg-Lyons, D., Keepers, K., Guo, Y., Bullard, J., Fierer, N., Carter, D. O., Knight, R., 2013. A microbial clock provides an accurate estimate of the postmortem interval in a mouse model system. *eLife* 2, e01104.

Mondor, E. B., Tremblay, M. N., Tomberlin, J. K., Benbow, E. M., Tarone, A. M., Crippen, T. L. 2012. The ecology of carrion decomposition. *Nat. Educ. Knowledge* 3:21.

NIH HMP Working Group, Peterson, J., Garges, S., Giovanni, M., McInnes, P., Wang, L., Schloss, J. A., Bonazzi, V., McEwen, J. E., Wetterstrand, K. A., Deal, C., Baker, C. C., Di Francesco, V., Howcroft, T. K., Karp, R. W., Lunsford, R. D., Wellington, C. R., Belachew, T., Wright, M., Giblin, C., David, H., Mills, M., Salomon, R., Mullins, C., Akolkar, B., Begg, L., Davis, C., Grandison, L., Humble, M., Khalsa, J., Little, A. R., Peavy, H., Pontzer, C., Portnoy, M., Sayre, M. H., Starke-Reed, P., Zakhari, S., Read, J., Watson, B., Guyer, M., 2009. The NIH Human Microbiome Project. *Genome Research* 19, 2317-2323.

Payne, J.,1965. A summer carrion study of the baby pig Sus scrofa Linnaeus. *Ecology* 46, 592-602.

Pechal, J. L., Crippen, T. L., Benbow, M. E., Tarone, A. M., Dowd, S., Tomberlin, J. K., 2014. The potential use of bacterial community succession in forensics as described by high throughput metagenomic sequencing. *International Journal of Legal Medicine* 128, 193-205.

Preece, P., Cairns, N. J., 2003. Quantifying mRNA in postmortem human brain: influence of gender, age at death, postmortem interval, brain pH, agonal state and inter-lobe mRNA variance. Brain research. *Molecular brain research* 118, 60-71.

Ribet, D., Cossart, P., 2015. How bacterial pathogens colonize their hosts and invade deeper tissues . *Microbes and Infection* 17, 173e183.

Sabucedo, A. J., Furton, K. G., 2003. Estimation of postmortem interval using the protein marker cardiac Troponin I. *Forensic science international* 134, 11-16.

Sampaio-Silva, F., Magalhaes, T., Carvalho, F., Dinis-Oliveira, R. J., Silvestre, R., 2013. Profiling of RNA degradation for estimation of post morterm interval. *PloS one* 8, e56507.

Schmitt, A., Cunha, E., Pinheiro, J. 2006. *Forensic Anthropology and Medicine*. Humana Press. New Jersey.

Schramm, M., Falkai, P., Tepest, R., Schneider-Axmann, T., Przkora, R., Waha, A., Pietsch, T., Bonte, W., Bayer, T. A., 1999. Stability of RNA transcripts in post-mortem psychiatric brains. *J. Neural Transm.* 106, 329-335.

Tibbett, M., Carter, D. 2008. Soil Analysis in forensic taphonomy. *Chemical and biological effects of buried human remains*. CRC Press. Boca Raton.

Tomberlin, J. K., Mohr, R., Benbow, M. E., Tarone, A. M., VanLaerhoven, S., 2011. A roadmap for bridging basic and applied research in forensic entomology. *Annual review of entomology* 56, 401-421.

Towne, E. G. (2000). Prairie vegetation and soil nutrient responses to ungulate car- casses. *Oecologia* 122, 232-239.

Trotter, S. A., Brill, L. B., 2nd, Bennett, J. P., Jr., 2002. Stability of gene expression in postmortem brain revealed by cDNA gene array analysis. *Brain research* 942, 120-123.

van Doorn, N. L., Wilson, A. S., Willerslev, E., Gilbert, M. T., 2011. Bone marrow and bone as a source for postmortem RNA. *Journal of forensic sciences* 56, 720-725.

Vass, A. A., 2011. The elusive universal post-mortem interval formula. *Forensic science international* 204, 34-40.

Vass, A. A., Barshick, S. A., Sega, G., Caton, J., Skeen, J. T., Love, J. C., Synstelien, J.A., 2002. Decomposition chemistry of human remains: a new methodology for determining the postmortem interval. *Journal of forensic sciences* 47, 542-553.

Vass A., 2001. Beyond the Grave - Understanding Human Decomposition. *Microbiology Today* 28: 190-192.

Vass, A. A., Bass, W. M., Wolt, J. D., Foss, J. E., Ammons, J. T., 1992. Time since death determinations of human cadavers using soil solution. *Journal of forensic sciences* 37, 1236-1253.

In: Forensic Science
Editor: Johanna Brewer
ISBN: 978-1-63483-086-7

Chapter 3

DIAGNOSIS OF DROWNING USING 16S RIBOSOMAL RNA GENE OF PHYTOPLANKTON

Jian Tie*[*] and Seisaku Uchigasaki
Division of Legal Medicine,
Nihon University School of Medicine,
Tokyo, Japan

ABSTRACT

In forensic medicine, the diagnosis of a corpse immersed in water that is differentiation between death from drowning or dead on entering the water, is made mainly using the diatom test by acid digestion. However, the diatom test by acid digesting is technically complicated and requires a larger quantity of sample. Moreover, the procedure is hazardous due to the use of strong acid and is time consuming. Alternative methods for the diagnosis of drowning by detecting plankton genes living in water using molecular biological technique have been reported. By performing one PCR amplification using various organs of a corpse immersed in water, this method is anticipated to permit rapid and accurate diagnosis of death from drowning by identifying the genes of phytoplankton that had entered solid organs via blood circulation. However, the currently used PCR method for the detection of

[*] Corresponding author: Jian Tie, Email: tetsu.ken@nihon-u.ac.jp

phytoplankton requires DNA purification, necessitating treatment of tissues as well as differentiation between sea water and fresh water phytoplankton. Thus this method has the disadvantages of being labor intensive and having the risk of contamination. To overcome these issues, first we targeted synechococcus sp (Bacteria; Cyanobacteria; Oscillatoriophycideae; Chroococcales), a picoplankton found in both sea water and fresh water, and designed primers specific for its 16S ribosomal RNA genes (16S rDNA). We succeeded to find that phytoplankton DNA used the quantity of the organization from the organ of the drowned body using a direct PCR method without DNA extraction to check the death from being drowned from autopsy findings and a diatom test by acid digestion and an autopsy case to be diagnosed. Furthermore, we quantitatively assayed the 16S rDNA gene of picoplankton from tissues of drowned rabbits and non-drowned rabbits immersed in water after death. The research was to investigate the quantities of picoplankton DNA in lung, liver, kidney tissues and blood in drowned and non-drowned rabbits, and the sensitivity of detection of picoplankton DNA by direct PCR for the diagnosis of death from drowning. The results verified that the detection of phytoplankton DNA in the liver and kidney is the most important evidence for the diagnosis of death from drowning.

INTRODUCTION

Diagnosis of the death by drowning from postmortem examinations is the most difficult in forensic medicine, although when a corpse immersed in water that is differentiation between death from drowning or dead on entering the water. At times, the doctors cannot identify the cause of death only in dissection views and often must usually give the last diagnosis by some kind of supporting inspection.

IMMERSION AND DROWNING

When a death is deemed suspicious or unexplained, medical examiners frequently perform autopsies to determine the exact cause. The manner of death is classified into one of five categories: (1) natural, (2) homicide, (3) suicide, (4) accident, or (5) undetermined, based on the circumstances. Immersion is not the same as drowning, but there a tendency to regard all dead bodies recovered from water as having drowned. Bodies recovered from water may have: (1) died of natural causes before entering water, such as a coronary victim slipping from the river bank, bridge or boat. (2) Died of natural causes in the water, a not uncommon event from ischemic heart disease, etc. (3) Died from exposure and hypothermia in the water, though drowning is usually the terminal mode of death. (4) Died of injuries or other unnatural cause before entering the water. (5) Died injuries after entering the water, such as striking rocks, bridges or being hit by boats or propellers. (6) Died from submersion, but not drowning; this includes 'shock' which is cardiac arrest due to cold water on the skin or laryngeal-pharyngeal area. (7) Died from true drowning after submersion, from aspiration of large volumes of water into the lungs. The mechanism of death in bodies who die as a result of submersion is not always classical drowning. In fresh water, fluid is hypotonic compared to plasma, so that when water pours into the lings, a rapid osmotic transfer takes place through the alveolar membranes. The blood volume may increase by 50 per cent within a minute, placing a great strain on the heart due to hypervolemia. In sea water, the fluid is hypertonic, so that water is withdrawn from the plasma into the lungs. In either type of drowning, the post-mortem findings, either on external observation or at autopsy may be variable. Drowning is one of the most difficult modes of death to prove at post-mortem, especially when the body is not examined in a fresh condition.

Common Diagnosis of Drowning Death

Classically, there will be a plume of froth at the mouth and nostrils, sometimes tinged with blood. Though externally this may be the only sign, it is often absent, especially when the body has been dead for some time. After autopsy there may found some change in the organs, but no special signs belong the drowning. The diatom test is controversial, having its opponents and champions. It was best described as suggestive and indicative technique, but is not sufficiently reliable to use as legal proof. The test depends on the presence of microscopic algae, called diatoms, which are present in sea water and most unpolluted natural fresh water. If a person drowns in diatom-containing water, the inhaled algae reach the lungs and some penetrate the alveolar walls. As the heart is still beating, these diatoms will be transported around the circulation and lodge in distant organs, such as the kidney, liver, brain and bone marrow. If at autopsy, samples of these tissues are taken with precautions against contamination from the body surface, the diatoms can be recovered by acid digestion and examination of a centrifuged deposit under the microscope. If a dead body is dropped into water, then although diatoms can reach the ling by passive percolation, no circulatory transfer can occur to other tissues and hence none can be discovered microscopically. Unfortunately, the test is often negative even in undoubted drowning and false positives are said to occur from a variety of technical reasons, as well the alleged presence of diatoms in non-drowned bodies. Therefore, the specialists in forensic medicine studied the diagnosis method of more certain drowning for a long time [1-2].

Diagnosis of Drowning Death Using Biomarkers

The biomarker is applied widely in not only the diagnosis of the drowning but also the field of other forensic sciences [3-4]. So far, the normal method of diagnosis drowning death is made mainly using morphology observation of the diatom test by acid digestion. However, the diatom test by acid digesting is technically complicated and requires a larger quantity of tissue samples. Moreover, the procedure is hazardous due to the use of strong acid, and is time consuming. Alternative methods for the diagnosis of drowning by detecting the DNA from phytoplankton or bacteria living in water using molecular biological technique have been reported [5-6]. The bacteria like Aeromonas hydrophila, Vibrionaceae, Photobacterium phosphoreum, P. leiognathi, Vibrio fisheri and V. harveyi are reported as a biomarker for identification of drowning death. However, there is it to not only river water and the seawater but also the soil, well water and a plant, breeding is fast, besides, and the bacteria are short of the accuracy of the diagnosis. As for the phytoplankton, it was studied a diagnosis of the drowning as not only the morphology but also biomarker. It is because there is phytoplankton in a river water and the sea water with a stable kind in comparison with bacteria. Normally, the phytoplankton is classified in picoplankton (2 um or less), nanoplankton (2-20 um) and microplankton (20-200 um) by size. Picoplankton belonging to the Synechococcus genus in cyanobacteria are found widely in river water and seawater.

Recently, several method of examining a corpse immersed in water using PCR to amplify specific phytoplankton genes has been reported [7]. By performing one PCR amplification using various organs of a corpse immersed in water, these methods are anticipated to permit rapid and accurate diagnosis of death from drowning by identifying the genes of phytoplankton that had entered solid organs via blood circulation. However, the currently used PCR method for the detection of phytoplankton requires DNA purification, necessitating treatment of tissues as well as differentiation between sea water and fresh water phytoplankton. Thus this method has the disadvantages of being labor intensive and having the risk of contamination. Therefore development of simpler and easier laboratory procedure is necessary.

DETECTED PHYTOPLANKTON FROM ORGANS OF DROWNING CORPSE USING DIRECT PCR

To overcome these issues, first we targeted synechococcus sp (Bacteria; Cyanobacteria; Oscillatoriophycideae; Chroococcales), a picoplankton found in both sea water and fresh water, and designed primers specific for its 16S ribosomal RNA genes (16S rDNA). Samples of human lung, liver and kidney tissues were collected from autopsied bodies in our laboratory. All the cases were diagnosed as death by drowning, from observations at autopsy and detection of diatoms from the tissues by the acid digestion method. Approximately 1 g each of lung, liver and kidney tissues were obtained from each body. Scrape samples of 1 and 5 mg were prepared from each tissue using a blade, and placed in 1.5-ml plastic tubes. Five hundred μl of digest buffer (pH 8.3) was added to each tube. After mixing the tissue samples with the digest buffer, the tubes was incubated in a microtube shaking incubator at 60ºC for 20 min, the tubes were placed in 100ºC for 3-5 min. The specific primer set used to amplify the 16S rDNA of phytoplankton in the present study was designed according to the sequence of the 16S rDNA for cyanobacteria provided by GenBank. The forward primer was 5'- CGTCTGATTAGCTAGTTGGT-3', and the reverse primer was 5'- CCCAATAATTCCGGATAACG-3'. PCR was performed in a 25 μl reaction mixture containing 1.0 μl sample of digested tissue template. The PCR products were subjected to electrophoresis in 5% polyacrylamide gels (C=3%) for 30 min. Then, the gels were stained with ethidium bromide solution and then photographed under transillumination. The PCR products of the digested tissue samples from drowned bodies yielded bands at 310 bp, corresponding to the 16S rDNA of cyanobacteria. The 16S rDNA fragment was amplified from all the lung, liver and kidney tissues of the drowned bodies, as well as from all the water samples collected at the scenes of drowning (Figure 1).

In the sensitivity study, approximately 4000 Euglena gracilis cells were required to isolate approximately 10 ng of phytoplankton DNA (Figure 2).

One mg each of the lung, liver and kidney tissue yielded at least 20 ng of plankton DNA in 500 μl of the digest buffer. E-PCR refers to a computational procedure used to search DNA sequences for sequence tagged sites (STS), each of which is defined by a pair of primer sequences and an expected PCR product size [8]. A total of 398 species of bacteria, fungi, plants, insects, and some mammals were searched by e-PCR in university of California santa cruz (UCSC) in silico PCR (http://mocrobes. ucsc. edu/index.html), national center for biotechnology information (NCBI) / BLAST (http://blast. nlm.nih. gov/) and other organisms in NCBI / uniSTS (http://www. ncbi nlm. Gov./ projects / e-pcr/). Only 5 species of cyanobacteria genome matched the PCR primers.

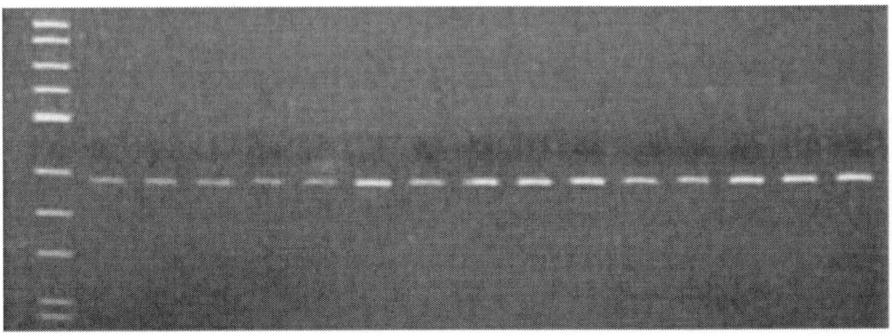

Figure 1. Polyacrylamide gel electrophoresis of PCR products for 16S rDNA amplified from human lung, liver and kidney tissues of drowned bodies. ϕX174DNA / Hinf I was used as DNA size marker.

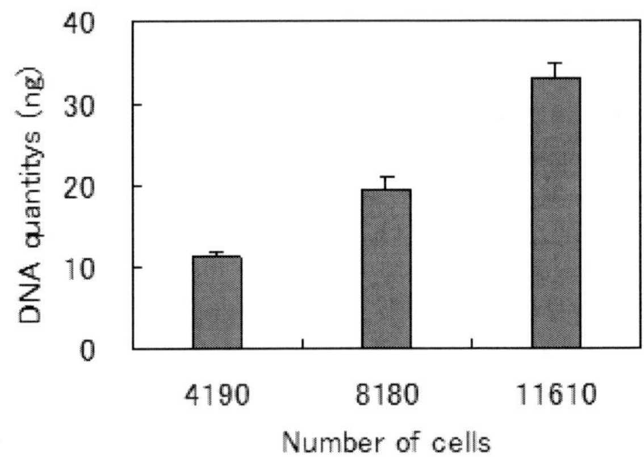

Figure 2. Quantities of DNA obtained from different numbers of Euglena gracilis cells.

The cyanobacteria are a widely distributed and diverse group of picoplankton that habitat both seawater and freshwater [9-11]. DNA sequence information of the 16S rDNA obtained from cyanobacteria is used to investigate the presence of cyanobacteria and their prevalence in natural habitats [12]. In some cases of drowning, diatoms could not be found by the acid digestion test in some tissues, but the picoplankton 16S rDNA was easily detected by PCR amplification in all tissues. The reason is that picoplankton are smaller than diatoms in size; thus picoplankton gain access into blood circulation easily and are found in internal organs. When we examined the effect of quantity of tissue on plankton detection by our PCR method, we found that the specific 16S rDNA fragment was clearly amplified when only 1 mg of scrape preparation of lung, liver or kidney tissue was used for the PCR amplification. When the amount of the scrape preparation is less than 5 mg, the tissue sample can be dissolved in the digest buffer. However when the amount of scrape preparation exceeds 5 mg, then the whole tissue sample may not be dissolved. Even in this case, there are enough templates in the digest for successful PCR amplification of the 16S rDNA.

Real-Time PCR Assay for Detection of Phytoplankton DNA Distribution in Tissues of Drowned Rabbits

A sensitive and specific method has developed for the diagnosis of drowned death using direct PCR of the human tissues from the samples that collected by autopsy. However, for the further establishment of this method, animal experiment is necessary, because we need to know whether the phytoplankton can be found from died of injuries or other unnatural cause before entering the water. For this reason, we quantitatively assayed the 16S rDNA of phytoplankton from tissues of drowned rabbits and non-drowned rabbits immersed in water after death. The aim of the research was to investigate the quantities of phytoplankton in lung, liver, kidney tissues and blood in drowned and non-drowned rabbits, and the sensitivity of detection of phytoplankton DNA by direct PCR for the diagnosis of death from drowning. Thirty mg each of liver, kidney and lung tissues and 100 μl blood were obtained from drowned rabbits and non-drowned rabbits. Plankton DNA in the tissues was extracted using the DNeasy® Blood & Tissue kit to determine the yield of plankton DNA from each tissue. TaqMan real-time PCR was performed for quantitative analysis of phytoplankton DNA. Target plankton DNA was detected in liver, kidney and lung samples obtained from the drowned rabbits, and the quantities of DNA was all over 200 ng. No phytoplankton DNA was detected in non-drowned rabbit tissues (except in lung samples), even though the bodies were immersed after death for three days. Using direct PCR without DNA extraction, phytoplankton DNA was amplified from approximately 5 mg of liver, kidney and lung tissues of all drowned rabbits as well as from lung tissue of some non-drowned rabbits, but not from liver or kidney tissue of non-drowned rabbits. The results obtained from this animal experiment verified that direct PCR for detection of phytoplankton DNA is useful for the diagnosis of drowning. Although we observed seasonal changes in quantity of planktons in river water, we were able to detect phytoplankton from various organs of drowned bodies during the season when planktons were not the most abundant.

EVOLUTION OF THE PCR METHOD FOR DIAGNOSIS OF DROWNING DEATH

Without DNA extraction, the direct PCR method was used to examine autopsy cases diagnosed as death from drowning and diatom test by acid digestion, and succeeded to detect phytoplankton DNA using minute quantities of tissues from organs of the drowned bodies. Furthermore, we quantitatively assayed the 16S rDNA gene of picoplankton from tissues of drowned rabbits and non-drowned rabbits immersed in water after death. The research was to investigate the quantities of picoplankton DNA in lung, liver, kidney tissues and blood in drowned and non-drowned rabbits, and the sensitivity of detection of picoplankton DNA by direct PCR for the diagnosis of death from drowning. No picoplankton DNA was detected in non-drowned rabbit tissues (except in lung samples), even though the bodies were immersed after death for three days. The results obtained from this animal experiment verified that direct PCR for detection of picoplankton DNA is useful for the diagnosis of drowning. Although we observed seasonal changes in quantity of planktons in river water, we were able to detect from various organs of drowned bodies during the season when planktons were not the most abundant.

Apart from autopsy findings, the detection of phytoplankton ingested with water and entered organs via blood circulation provides the most important evidence for a diagnosis of death from drowning [13]. Compared with the conventional diatom test by acid digestion, the diagnostic method of detecting plankton DNA is a simple procedure that is capable of detecting planktons with high efficiency from a small quantity of sample. However, while the PCR method increases the sensitivity of detection compared to the conventional acid digestion method, it is necessary to verify that the PCR method also detects no phytoplankton in organs of corpses already dead on entering water. In the present study, we conducted animal experiments to quantify plankton DNA in solid organs (lung, liver and kidney) commonly used in forensic medicine for the diagnosis of drowning, and compared the quantities of DNA in the case of drowned death and the case of non-drowned death.

Sometimes, the diagnosis of the drowning is necessary for the confirmation of the place of the drowning other than differentiation of the death by drowning with drowning in some cases posthumously, in this case, it can distinguish with specific sequence of 16S rDNA collected from different phytoplankton in the river water or the seawater.

CONCLUSION

In conclusion, our method allows direct amplification of picoplankton DNA from small quantities of tissue fragments, and is useful for rapid diagnosis of death by drowning. We reported the first using the direct PCR for the diagnosis of drowned death, the data all obtained from human tissues of autopsies. Therefore, the animal experiment was performed for quantification of picoplankton from the tissues and the results verified that the detection of phytoplankton DNA in the liver and kidney is the most important evidence for the diagnosis of death from drowning. When phytoplankton DNA is amplified only from the lungs, the result must be interpreted very cautiously regarding the diagnosis of death by drowning. In addition, using the direct PCR, despite the seasonal changes in abundance of phytoplankton in river water, we were able to detect phytoplankton from various organs during the season when phytoplankton was not the most abundant.

Acknowledgments

We would like to thank the members of our laboratory Eiji Isobe and Isamu Isahai as well as Takeshi Haseba, Fanlai Cui and Youkichi Ohno (Nippon Medical School) who supported this study.

References

[1] Ludes B., et al., Continuous river monitoring of the diatoms in the diagnosis of drowning. *J. Forensic. Sci.,* 1996. 41(3): p. 425-8.

[2] Hürlimann J., et al., Diatom detection in the diagnosis of death by drowning. *Int. J. Legal Med.*, 2000. 114(1-2): p. 6-14.

[3] Aoyagi M., et al., A novel method for the diagnosis of drowning by detection of Aeromonas sobria with PCR method. *Leg. Med.*, 2009. 11(6): p. 257-9.

[4] Takeichi T. and Kitamura O., Detection of diatom in formalin-fixed tissue by proteinase K digestion. *Forensic. Sci. Int.*, 2009. 190(1-3): p. 19-23.

[5] Kobayashi M., et al., Novel detection of plankton from lung tissue by enzymatic digestion method. *Forensic. Sci. Int.*, 1993. 60(1-2): p. 81-90.

[6] Suto M., et al., PCR detection of bacterial genes provides evidence of death by drowning. *Leg. Med.,* 2009. Supply 1 S: p. 354-6.

[7] Suto M., et al., Phytoplankton gene detection in drowned rabbits. *Leg. Med.,* 2003. Supply 1S: p. 142-4.

[8] Rotmistrovsky K., et al., A web server for performing electronic PCR. *Nucleic. Acids Res.*, 2004. 32: p. 108-12.

[9] Rahman M.A., et al., Toxicity of arsenic species to three freshwater organisms and biotransformation of inorganic arsenic by freshwater phytoplankton (Chlorella sp. CE-35). *Ecotoxicol. Environ. Saf.*, 2014. 106: p. 126-35.

[10] Saniewska D., et al., Mercury loads into the sea associated with extreme flood. *Environ. Pollut.* 2014. 191: p. 93-100.

[11] Chekalyuk A. and Hafez M., Analysis of spectral excitation for measurements of fluorescence constituents in natural waters. *Opt. Express.* 2013. 21(24): p. 29255-68.

[12] Rudi K., et al., Application of sequence-specific labeled 16S rRNA gene oligonucleotide probes for genetic profiling of cyanobacterial abundance and diversity by array hybridization. *Appl. Environ. Microbiol.*, 2000. 66(9): p. 4004-11.

[13] Qu J. and Wang E., A study on the diagnosis of drowning by examination of lung chlorophyll (a) of planktons with a spectrofluorophotometer. *Forensic. Sci. Int.*, 1992. 53(2): p. 149-55.

BIBLIOGRAPHY

A dictionary of forensic science LCCN: 2011942657 Type of material: Book Personal name: Bell, Suzanne. Main title: A dictionary of forensic science / Suzanne Bell. Published/Created: Oxford; New York: Oxford University Press, c2012. Description: 303 p.: ill.; 20 cm. ISBN: 9780199594009 (pbk.) 0199594007 (pbk.) LC classification: HV8073.B4246 2012 Cover title: Oxford dictionary of forensic science Subjects: Forensic sciences--Dictionaries. Notes: Includes bibliographical references (p. 299-300). Series: Oxford paperback reference Dewey class no.: 363.2503

A hands-on introduction to forensic science: cracking the case LCCN: 2014025689 Type of material: Book Personal name: Okuda, Mark. Main title: A hands-on introduction to forensic science: cracking the case / Mark Okuda, Frank H. Stephenson, PhD. Published/Produced: Boca Raton: CRC Press, 2014. Projected pub date: 1111 Description: pages cm Links: Cover image http://images.tandf.co.uk/common/jackets/websmall/978148223/9781482234909.jpg ISBN: 9781482234909 (hardback) LC classification: HV8073.O453 2014 Related names: Stephenson, Frank H. Summary: "One failing of many forensic science textbooks is the isolation of chapters into compartmentalized units. This format prevents students from understanding the connection between material learned in previous chapters with that of the current chapter. Using a unique format, A Hands-On Introduction to Forensic

Science: Cracking the Case approaches the topic of forensic science from a real-life perspective in a way that these vital connections are encouraged and established.The book utilizes an ongoing fictional narrative throughout, entertaining students as it provides hands-on learning in order to "crack the case." As two investigators try to solve a missing persons case, each succeeding chapter reveals new characters, new information, and new physical evidence to be processed. A full range of topics are covered, including processing the crime scene, lifting prints, trace and blood evidence, DNA and mtDNA sequencing, ballistics, skeletal remains, and court testimony. Following the storyline, students are introduced to the appropriate science necessary to process the physical evidence, including math, physics, chemistry, and biology. The final element of each chapter includes a series of cost-effective, field-tested lab activities that train students in processing, analyzing, and documenting the physical evidence revealed in the narrative. Practical and realistic in its approach, this book enables students to understand how forensic science operates in the real world" -- Provided by publisher. "Preface The aim of this book is to present a novel way of teaching forensic science and, more importantly, a different way of presenting any science discipline whether it be chemistry, physics, or biology. Most science textbooks today are formatted and presented in a compartmentalized manner; students are likely to encounter a chapter entitled "Cell Structure and Function" in a biology textbook, or "States of Matter" in a chemistry textbook, or "Newton's First Law of Motion-Inertia" in a physics textbook, or "Blood Spatter" in a forensic textbook. These compartmentalized units present an extensive list of vocabulary words as well as scientific principles and concepts to be mastered. The student take notes, conducts lab experiments and, at the conclusion of the unit, is evaluated by some form of assessment. This cycle repeats itself with the next chapter until the course is completed at the end of the academic year. This format often lacks the connections learned in prior chapters to the current chapter under study. Science on the other hand is about making connections. The disconnect that commonly exists between science textbooks, science courses and the way real science operates will be addressed in this manuscript. Science works with a problem or an observation which requires an integrated approach involving many different disciplines of science coming together and the connections that these disciplines can provide to help solve a problem or explain an observation. This book attempts to capture the student's interest by means of a unique format. It begins with a story about two characters who go about solving a missing persons case"-- Provided by publisher. Subjects: Criminal investigation. Forensic ballistics. Chemistry, Forensic. LAW / Criminal Law / General. LAW / Forensic Science. Notes: Includes bibliographical references and index. Dewey class no.: 363.25

Bayesian networks for probabilistic inference and decision analysis in forensic science LCCN: 2014014247 Type of material: Book Personal name: Taroni, Franco, author. Main title: Bayesian networks for probabilistic inference and decision analysis in forensic science / Franco Taroni, Alex Biedermann, Silvia Bozza, Paolo Garbolino, Colin Aitken. Edition: Second edition. Published/Produced: Chichester, West Sussex; Hoboken, NJ: Wiley, 2014. Projected pub date: 1410 Description: pages cm Links: Cover image http://catalogimages.wiley.com/images/db/jimages/9780470979730.jpg ISBN: 9780470979730 (hardback) LC classification: QA279.5.T37 2014 Summary: "The aim is to offer theoretical and practical elements to help solve the following questions"-- Provided by publisher. Contents: Machine generated contents note: Preface to the first edition Preface to the second edition 1 The logic of decision 1.1 Uncertainty and probability 1.1.1 Probability is not about numbers, it is about coherent reasoning under uncertainty 1.1.2 The first two laws of probability 1.1.3 Relevance and independence 1.1.4 The third law of probability 1.1.5 Extension of the conversation 1.1.6 Bayes' theorem 1.1.7 Probability trees 1.1.8 Likelihood and probability 1.1.9 The calculus of (probable) truths 1.2 Reasoning under uncertainty 1.2.1 The Hound of the Baskervilles 1.2.2 Combination of background information and evidence 1.2.3 The odds form of Bayes' theorem 1.2.4 Combination of evidence 1.2.5

likelihood ratios 10.3.4 Construction of a Bayesian network 10.4 Pre-assessment in a cross-transfer scenario 10.4.1 Bi-directional transfer 10.4.2 A Bayesian network for a pre-assessment of a cross-transfer scenario 10.4.3 The expected value of the findings 10.5 Pre-assessment for consignment inspection 10.5.1 Inspecting small consignments 10.5.2 Bayesian network for inference about small consignments 10.5.3 Pre-assessment for inspection of small consignments 10.6 Pre-assessment for gunshot residue particles 10.6.1 Formation and deposition of gunshot residue particles 10.6.2 Bayesian network for grouped expected findings (GSR counts) 10.6.3 Examples for GSR count pre-assessment using a Bayesian network 11 Bayesian decision networks 11.1 Decision making in forensic science 11.2 Examples of forensic decision analyses 11.2.1 Deciding about whether or not to perform a DNA analysis 11.2.2 Probability assignment as a question of decision making 11.2.3 Decision analysis for consignment inspection 11.2.4 Decision after database searching 11.3 Further readings 12 Object oriented networks 12.1 Object-orientation 12.2 General elements of object oriented networks 12.2.1 Static versus dynamic networks 12.2.2 Dynamic Bayesian networks as object oriented networks 12.2.3 Refining internal class descriptions 12.3 Object oriented networks for evaluating DNA profiling results 12.3.1 Basic disputed paternity case 12.3.2 Useful class networks for modeling kinship analyses 12.3.3 Object oriented networks for kinship analyses 12.3.4 Object oriented networks for inference of source 12.3.5 Refining internal class descriptions and further considerations 13 Qualitative, sensitivity and conflict analyses 13.1 Qualitative probability models 13.1.1 Qualitative influence 13.1.2 Additive synergy 13.1.3 Product synergy 13.1.4 Properties of qualitative relationships 13.1.5 Implications of qualitative graphical models 13.2 Sensitivity analyses 13.2.1 Preliminaries 13.2.2 Sensitivity to a single probability assignment 13.2.3 Sensitivity to two probability assignments 13.2.4 Sensitivity to prior distribution 13.3 Conflict analysis 13.3.1 Conflict detection 13.3.2 Tracing a conflict 13.3.3 Conflict resolution Bibliography Author Index Subject Index. Subjects: Bayesian statistical decision theory--Graphic methods. Uncertainty (Information theory)--Graphic methods. Forensic sciences--Graphic methods. MATHEMATICS / Probability & Statistics / General. Notes: Includes bibliographical references and index. Additional formats: Online version: Taroni, Franco, author. Bayesian networks for probabilistic inference and decision analysis in forensic science Second edition. Chichester, West Sussex; Hoboken, NJ: Wiley, 2014 9781118914748 (DLC) 2014018080 Dewey class no.: 363.2501/519542

Careers in chemistry LCCN: 2012044523 Type of material: Book Main title: Careers in chemistry / editor, Donald R. Franceschetti, The University of Memphis. Published/Produced: Ipswich, Massachusetts: Salem Press, a division of EBSCO Publishing; Amenia, NY: Grey House Publishing, 2013. Description: xxi, 482 pages; 24 cm ISBN: 9781587659935 (hardcover) 158765993X (hardcover) LC classification: QD39.5.C243 2013 Related names: Franceschetti, Donald R., 1947- editor of compilation. Contents: Agricultural chemistry -- Agricultural chemist -- Atmospheric chemistry -- Atmospheric chemist -- Biochemistry -- Biochemical engineer -- Chemical engineering -- Production engineer -- Chemical genetics -- Geneticist -- Chemical technology -- Chemical technician -- Cosmetics chemistry -- Perfume chemist -- Electrochemistry -- Fuel cell engineer -- Environmental chemistry -- Environmental chemist -- Femtochemistry -- Experimental scientist -- Food chemistry -- Food technologist or nutritionist -- Forensic science -- Forensic scientist -- Geochemistry -- Geochemist -- Green chemistry -- Analytical chemist -- Materials science -- Materials chemist -- Medicinal chemistry -- Oncologist -- Nuclear chemistry -- Radiochemist -- Petrochemistry -- Petrochemist -- Pharmacology -- Pharmacist -- Polymer chemistry -- Plastics engineer -- Pulp and paper chemistry -- Wood chemist -- Toxicology -- Toxicologist. Subjects: Chemistry--Vocational guidance. Notes: Includes bibliographical references (pages 441-450) and index. Dewey class no.: 540.23

Careers in criminal justice LCCN: 2013427585 Type of material:

Book Personal name: Johnston, Coy H., author. Main title: Careers in criminal justice / Coy H. Johnston. Published/Produced: Los Angeles, California: Sage Publications, Inc., 2015 ©2015 Description: xiv, 229 pages: illustrations; 26 cm ISBN: 9781483331461 (paperback) 1483331466 (paperback) LC classification: HV8143.J642 2015 Variant title: Criminal justice Summary: Overview: Coy H. Johnston's concise, yet thorough and creative text, Careers in Criminal Justice, prepares students to plan, pursue, and realize their career goals - from conception through the hiring process. The book's contemporary approach emphasizes student self-reflection and pragmatism in the pursuit of self-fulfillment and happiness. With coverage of over forty careers in policing, courts, corrections, and victims services, this text provides a comprehensive overview of the most popular and growing careers in the field. Readers are provided self assessment tools to enhance self awareness and steer them toward realistic and suitable careers in criminal justice. This easy-to-read guide is organized to facilitate understanding and encourage growth throughout the reader's career. Contents: Preface -- Acknowledgments -- About the author -- Section 1: Planning: -- Know Yourself: -- Introduction -- What you don't know can hurt you: -- What you don't know -- Self-assessments: -- What color is your parachute? -- Keirsey temperament sorter -- True colors -- What type am I? -- Four temperaments -- What color is your personality? -- Jung typology test -- GAcollege411 -- Summary -- Discussion questions -- Setting Realistic Goals: -- Introduction -- Set realistic goals -- Issues with police work: -- Use of deception -- Role of acting in police work -- Use of discretion -- Use of force -- Polygraph test -- Interviewing and interrogating -- Internal investigations -- Issues in the judicial system: -- Power of persuasion -- Attitude -- Communication skills -- Salary issues -- Power and responsibility -- Ethics in court -- Battle of wits -- Issues in Corrections: -- Communication gap -- Dangers for Correctional Officers -- Alternative Correctional Careers -- Realities of working in Criminal Justice -- Change -- Search for happiness -- Summary -- Discussion questions -- Section 2: Pursuing: -- Local And State Law

Enforcement: -- Introduction -- Local Law Enforcement -- Guest Speaker: Jan Strauss, Retired Police Chief -- Game And Fish Warden -- Gaming Officer -- Campus Police -- Nonsworn Opportunities -- Diversity issues -- Summary -- Discussion questions -- Federal Law Enforcement -- Introduction -- FBI -- Guest Speaker: Steve Palmer, FBI Special Agent -- DEA -- ATF -- Border Patrol -- ICE -- U S Secret Service -- USMS -- Nonsworn Positions -- Diversity in Federal Law Enforcement -- Experience -- Summary -- Discussion questions -- Forensic Science: -- Introduction -- CST -- Guest Speaker: Shayleen Eastman, Crime Scene Specialist III -- Forensic Scientist -- Medical examiner -- Investigative Technician -- Criminalist: -- DNA Analyst -- Toxicologist -- Forensic Nurse examiner -- Forensic Psychiatrist Specialist -- Forensic Interviewer -- Forensic Entomologist -- Diversity in Forensic Jobs -- Summary -- Discussion questions -- Victim Services: -- Introduction -- Victim Advocate and related positions -- Guest Speaker: Gina McAlister, Victim Services Specialist -- Diversity in Victim Services -- Summary -- Discussion questions -- Courts: -- Introduction -- Law School -- Job search -- Law Careers -- Guest Speaker: Charles W Ervin, Superior Court Judge -- Diversity in Law Careers -- Summary -- Discussion questions -- Corrections: -- Introduction -- State Correctional Officer -- County Detention Officer -- City Jailer -- Guest Speaker: Julie Glover, Sergeant, San Diego Sheriff's Department -- Parole and Probation Officers -- Juvenile Rehabilitation Counselor -- Home Detention Officer -- Probation Aide -- Parole Board Hearing Officer -- Diversity in Probation and Parole Work -- Summary -- Discussion questions -- Private Security And Investigations: -- Introduction -- Security Officer -- Guest Speaker: Timothy W Newton, Private Security Supervisor -- Private Investigator -- Loss Prevention -- Armed Security Guard -- Diversity issues in Private Security -- Summary -- Discussion questions -- Section 3: Preparing: -- Getting A Foot In The Door: -- Introduction -- Portfolios: -- Letters of recommendation -- Certificates -- Volunteer service -- Education documentation -- CV and resume -- Guest Speaker: Lance

Larsen, Juvenile Detention Officer -- Interview preparation: -- Police ride-along -- Court and attorney observations -- Firearms -- Physical fitness -- Volunteering versus internships -- Education and grades -- LSAT -- Application -- Written tests -- Psychological tests -- Summary -- Discussion questions -- Interview: -- Introduction -- Guest Speaker: Michael Norzagaray, Former Juvenile Probation Officer -- Punctuality -- Appearance -- Body language -- Introductions -- Sample questions -- Dos and don'ts -- Personal note to minorities, women, gays, and lesbians -- Note to law students -- Summary -- Index.. Subjects: Criminal justice, Administration of--Vocational guidance. Law enforcement--Vocational guidance--United States. Police--Vocational guidance--United States. Job hunting--United States. Notes: Includes bibliographical references and index. Dewey class no.: 363.2/023/73

Careers in forensic science LCCN: 2011045887 Type of material: Book Personal name: Woog, Adam, 1953- Main title: Careers in forensic science / Adam Woog. Published/Created: New York: Marshall Cavendish, c2013. Projected pub date: 1309 Description: p. cm. ISBN: 9781608709588 (print) 9781608709656 (ebook) LC classification: HV8073.8.W665 2013 Summary: "Provides a comprehensive look at the careers available within different fields of the forensic science and how students can plan toward a career in a forensic science"--Provided by publisher. Contents: The big picture -- What it takes: requirements -- Higher education and training -- On the job -- Salaries and benefits. Subjects: Forensic sciences--Vocational guidance--Juvenile literature. Criminal investigation--Vocational guidance--Juvenile literature. Notes: Includes bibliographical references and index. Series: Law and order jobs Dewey class no.: 363.25023

Careers in forensic science LCCN: 2013030140 Type of material: Book Personal name: Woog, Adam, 1953- Main title: Careers in forensic science / by Adam Woog. Published/Produced: New York: Cavendish Square, 2014. Description: 112 pages; 24 cm. ISBN: 9781627124195 (library binding) 9781627124201 (pbk.) LC classification: HV8073.W66 2014 Contents: Introduction --

The big picture -- What it takes -- Higher education and training -- On the job -- Salaries and benefits -- Glossary -- Bibliography -- For further research -- Notes -- About the author. Subjects: Forensic sciences--Vocational guidance. Criminal investigation--Vocational guidance. Notes: Includes bibliographical references (page 107) and index. Series: Law and order jobs Dewey class no.: 363.25023

Conviction of the innocent: lessons from psychological research LCCN: 2011021401 Type of material: Book Main title: Conviction of the innocent: lessons from psychological research / edited by Brian L. Cutler. Edition: 1st ed. Published/Created: Washington, DC: American Psychological Association, c2012. Description: xiii, 370 p.; 27 cm. ISBN: 9781433810213 1433810212 LC classification: HV7419.C692 2012 Related names: Cutler, Brian L. Contents: I. The suspects -- 1. At-risk populations under investigation and at trial / Robert J. Norris and Allison D. Redlich -- II. The investigators -- 2. Detecting deception / Amy-May Leach -- 3. False confessions / Lisa E. Hasel and Saul M. Kassin -- 4. Procedural justice evaluations in interrogations / Diane Sivasubramaniam and Larry Heuer -- III. The eyewitnesses -- 5. Fallible eyewitness memory and identification / Kathy Pezdek -- 6. Suggestive eyewitness identification procedures / David M. Zimmerman, Jacqueline L. Austin, and Margaret Bull Kovera -- 7. Eyewitness confidence malleability / Amy Bradfield Douglass and Afton Pavletic -- 8. Why do motions to suppress suggestive eyewitness identifications fail? / Gary L. Wells, Sarah M. Greathouse, and Laura Smalarz -- 9. Jurors believe eyewitnesses / Carolyn Semmler, Neil Brewer, and Amy Bradfield Douglass -- IV. The trial witnesses -- 10. Unreliable informant testimony / Jeffrey S. Neuschatz... [et al.] -- 11. Alibi witnesses / Tara M. Burke and Stéphanie B. Marion -- 12. Pscyhological perspectives on problems with forensic science evidence / Itiel E. Dror and Rebecca Bucht -- V. Pervasive issues -- 13. Race and racism / Ellen S. Cohn, Donald Bucolo, and Samuel R. Sommers -- 14. Tunnel vision / Keith A. Findley -- VI. The exonerated -- 15. Life after wrongful conviction / Kimberley A. Clow, Amy-May Leach, and Rosemary

Ricciardelli -- VII. Conclusion -- Conclusion / Brian L. Cutler. Subjects: Criminal justice, Administration of--Psychological aspects. Criminal investigation--Psychological aspects. Notes: Includes bibliographical references and index. Dewey class no.: 364.3

Cops in lab coats: curbing wrongful convictions through independent forensic laboratories LCCN: 2014044884 Type of material: Book Personal name: Thompson, Sandra Guerra. Main title: Cops in lab coats: curbing wrongful convictions through independent forensic laboratories / Sandra Guerra Thompson. Published/Produced: Durham: Carolina Academic Press, 2015. Projected pub date: 1111 Description: pages cm ISBN: 9781611635294 (alk. paper) LC classification: HV6795.H6 T46 2015 Contents: Foreword -- Acknowledgments -- Glossary -- Bad science wreaks havoc-the George Rodriguez story -- A national crisis -- The complex critique of forensic science -- The paradox of the ethical-and biased--analyst -- Safeguarding justice -- Resisting independence -- Houston's laboratory experiment -- Conclusion: reimagining the practice of forensic science -- Epilogue -- Bibliography -- Index -- About the author. Subjects: Criminal investigation--Texas--Houston. Forensic sciences--Texas--Houston. Judicial error--Texas--Houston. Notes: Includes bibliographical references and index. Dewey class no.: 363.25/60973

Criminal justice and forensic science: a multidisciplinary introduction LCCN: 2014038021 Type of material: Book Personal name: Smith, Lisa L. (Criminologist) Main title: Criminal justice and forensic science: a multidisciplinary introduction / Lisa Smith, John Bond. Published/Produced: Basingstoke, Hampshire; New York, NY: Palgrave Macmillan, 2014. Projected pub date: 1412 Description: pages cm ISBN: 9781137310255 (paperback) LC classification: HV8073.S6244 2014 Contents: A multidisciplinary approach to forensic studies -- Identification evidence -- Comparison evidence -- Crime scene examination -- Interpreting evidence -- Forensic science in the popular media -- Forensic science in the courtroom -- Issues in contemporary forensic science and justice. Subjects: Forensic sciences. Criminal justice, Administration of. Notes: Includes bibliographical references and index. Dewey class no.: 363.25.

Criminalistics: an introduction to forensic science LCCN: 2013045701 Type of material: Book Personal name: Saferstein, Richard, 1941- Main title: Criminalistics: an introduction to forensic science / Richard Saferstein, Ph.D., Forensic Science Consultant, Mt. Laurel, New Jersey. Edition: Eleventh Edition Published/Produced: Boston: Pearson, [2015] Description: xvii, 529 pages: color illustrations; 27 cm ISBN: 9780133458824 0133458822 LC classification: HV8073.S2 2015 Contents: Introduction -- The crime scene -- Physical evidence -- Crime-scene reconstruction: bloodstain pattern analysis -- Death investigation -- Fingerprints -- The microscope -- Firearms, tool marks, and other impressions -- Matter, light, and glass examination -- Hairs and fibers -- Drugs -- Forensic toxicology -- Metals, paint, and soil -- Forensic serology -- Dna: the indispensable forensic science tool -- Forensic aspects of fire and explosion investigation -- Document examination -- Computer forensics -- Mobile device forensics. Subjects: Criminal investigation. Forensic ballistics. Chemistry, Forensic. Medical jurisprudence. Notes: Includes bibliographical references and index. Dewey class no.: 363.25

Criminalistics: forensic science, crime, and terrorism LCCN: 2013034072 Type of material: Book Personal name: Girard, James, author. Main title: Criminalistics: forensic science, crime, and terrorism / James E. Girard, American University, Department of Chemistry, Washington, DC. Edition: Third edition. Published/Produced: Burlington, MA: Jones & Bartlett Learning, [2015] Description: xii, 525 pages: chiefly color illustrations; 29 cm ISBN: 9781449691790 (casebound) 144969179X (casebound) 9781284037036 1284037037 LC classification: HV8073.G564 2015 Subjects: Forensic sciences. Criminal investigation. Notes: Includes bibliographical references and index. Dewey class no.: 363.25/62

Developments in handwriting and signature identification in the digital age LCCN: 2012537249 Type of material: Book Personal name: Harralson, Heidi H. Main title: Developments in handwriting and signature identification in the digital age / Heidi H. Harralson; series editor, Larry S. Miller. Published/Created: Oxford: Anderson, 2013. Description: viii, 140 p.: ill.; 23 cm. ISBN: 9781455731473 (pbk) 1455731471 (pbk) Related names: Miller, Larry, 1953 August 26- Summary: Consists of short-format content on developments, unique perspectives, or how-to information on areas in forensic science - all specifically designed to meet the needs of the criminal justice community. This title opens up the world of forensic science to the criminal justice community. Contents: Machine generated contents note: ch. 1 Introduction -- 1.1.Key Concepts in Forensic Handwriting Examination -- 1.2.Forgery and Simulation -- ch. 2 The Decline of Handwriting -- 2.1.The Need for Handwriting Instruction -- 2.2.The Importance of Handwriting -- 2.3.Handwriting Instruction Methods -- 2.4.Influence of Learning Disabilities on Handwriting -- 2.5.Forensic Problems with Illegible and Printed Handwriting -- 2.6.Summary -- ch. 3 Advances in Handwriting Research and Technology -- 3.1.Handwriting and Movement Disorders -- 3.2.Handwriting and Psychiatric Disorders -- 3.3.Other Factors Influencing Handwriting -- 3.4.Handwriting and Simulation -- 3.5.Handwriting and Disguise -- 3.6.Handwriting and Neuro-technology -- 3.7.Summary -- ch. 4 Digital and Electronic Handwriting -- 4.1.The Electronic Signature Act -- 4.2.Applications for Electronic Signatures -- 4.3.Security and Privacy Issues -- 4.4.Digital Technology -- 4.5.Digital and Electronic Signatures -- Contents note continued: 4.6.Digitization of Signatures: Optical Scanning and Facsimile -- 4.7.Digital Tablets -- 4.8.The Stylus and Other Writing "Instruments" -- 4.9.Handwriting Recognition Technology -- 4.10.Automated Forensic Handwriting Analysis Technology -- 4.11.Standardization -- 4.12.Summary -- ch. 5 Forensic Analysis of Electronic Signatures -- 5.1.Practical Application -- 5.2.E-signature Signing Process and Standardization -- 5.3.Technological Considerations -- 5.4.Evaluating Static and Dynamic Handwriting Features -- 5.5.Forgery and

System Attacks -- 5.6.Limitations -- 5.7.Legal Implications -- 5.8.Proposed Methodology -- 5.9.Summary -- ch. 6 The Law, Science, and Handwriting Identification -- 6.1.Forensic Science Reform -- 6.2.A Critical Review of Forensic Handwriting Examination -- 6.3.Status of Forensic Handwriting Identification -- 6.4.Education and Training -- 6.5.Certification and Laboratory Accreditation -- 6.6.Proficiency Testing -- Contents note continued: 6.7.Scientific Validity and Reliability -- 6.8.Terminology and Methodology -- 6.9.Advances in Handwriting Identification Technology -- 6.10.Legal Challenges to Scientific Admissibility -- 6.11.Summary. Subjects: Writing--Identification. Graphology. Notes: Includes bibliographical references. Library of Congress Holdings Information not available.

Digging the disappeared: forensic science after atrocity LCCN: 2014039062 Type of material: Book Personal name: Rosenblatt, Adam (Adam Richard), author. Main title: Digging the disappeared: forensic science after atrocity / Adam Rosenblatt. Published/Produced: Stanford, California: Stanford University Press, 2015. Projected pub date: 1503 Description: pages cm ISBN: 9780804788779 (cloth: alk. paper) 9780804794916 (pbk.: alk. paper) LC classification: GN69.8.R67 2015 Contents: Introduction: born at the graves: a human rights movement takes shape -- The stakeholders in international forensic investigations -- The politics of grief -- Forensics of the sacred -- Dead to rights -- Caring for the dead. Subjects: Forensic anthropology--Moral and ethical aspects. Dead--Identification. Mass burials. Human rights. Atrocities. Notes: Includes bibliographical references and index. Dewey class no.: 599.9

DNA analysis LCCN: 2013006934 Type of material: Book Personal name: Hunter, William, 1971- Main title: DNA analysis / William Hunter. Published/Created: Broomall, PA: Mason Crest, c2014. Description: 112 p.: col. ill.; 25 cm. ISBN: 9781422228647 (hardcover) 9781422228616 (series) 9781422289501 (ebook) LC classification: RA1057.5.H862 2014 Contents: The changing face of forensic science -- Introducing DNA! --

DNA profiling: a beginner's guide -- Evidence and accuracy -- The use of DNA databases to catch criminals -- Tools of the trade -- Forensic biology: a promising career? Subjects: Forensic genetics--Juvenile literature. DNA fingerprinting--Juvenile literature. DNA--Analysis--Juvenile literature. Medical jurisprudence--Juvenile literature. Forensic sciences--Juvenile literature. Notes: Includes bibliographical references (p. 106-108) and index. 012. Grades 7 to 8. Series: Solving crimes with science: forensics Dewey class no.: 614.1

Ethics in forensic science LCCN: 2011942946 Type of material: Book Main title: Ethics in forensic science / edited by J.C. Upshaw Downs, Anjali Ranadive Swienton. Published/Created: Oxford; Waltham, Mass.: Academic, 2012. Description: xiii, 454 p.: ill.; 24 cm. ISBN: 9780123850195 (hbk.) 0123850193 (hbk.) LC classification: HV8073.E84 2012 Related names: Downs, J. C. Upshaw. Swienton, Anjali Ranadive. Contents: A brief history of ethics / Jennifer Downs -- The forensic practitioner's quest for truth / Fr. William Willoughby III, Elena Thompson, and J.C. Upshaw Downs -- General forensic ethical dilemmas / Jay A. Siegel -- Codes of ethics in forensic science societies: the organizational parameters of morality and conduct / Kenneth E. Melson -- Ethical foundations for SANE practice / Linda E. Ledray and Susan Chasson -- Ethics codes in other organizations: structures and enforcement / J.C. Upshaw Downs and Anjali Ranadive Swienton -- Teaching ethics / Max M. Houck and Robin T. Bowen -- Ethics in law enforcement: defining the thin blue line / Fred E. Stephens and J.C. Upshaw Downs -- Criminal investigational ethics / James Markey -- Ethics in the crime laboratory and in crime scene investigations / Barry A.J. Fisher -- Ethics for child abuse prosecutors: a call to leadership / Rachelle L. Carnesale -- Ethical duties of attorneys and experts in cases involving forensic evidence, from the perspective of the defense / Andrea Roth -- Ethics in testimony / Joseph P. Bono, Anjali Ranadive Swienton, and J.C. Upshaw Downs -- Judicial ethics / Haskell M. Pitluck -- The post-conviction process - from the perspective of the convicted /

Anjali Ranadive Swienton and J.C. Upshaw Downs -- Whistleblowers / J.C. Upshaw Downs -- Ethics in media / Anjali Ranadive Swienton and J.C. Upshaw Downs. Subjects: Forensic sciences--Moral and ethical aspects. Notes: Includes bibliographical references and index. Dewey class no.: 174.936325

Firearms, the law, and forensic ballistics LCCN: 2011041793 Type of material: Book Personal name: Warlow, T. A. (Tom A.) Main title: Firearms, the law, and forensic ballistics / Tom Warlow. Edition: 3rd ed. Published/Created: Boca Raton, FL: CRC Press, c2012. Description: xix, 477 p., [16] p. of plates: ill. (some col.); 25 cm. ISBN: 9781439818275 (hardcover: alk. paper) 1439818274 (hardcover: alk. paper) LC classification: HV8077.W37 2012 Subjects: Forensic ballistics. Forensic ballistics--Great Britain. Firearms--Law and legislation--Great Britain. Notes: Includes bibliographical references and index. Series: International forensic science and investigation series International forensic science and investigation series. Dewey class no.: 363.25/62

Fisher's techniques of crime scene investigation LCCN: 2012021097 Type of material: Book Personal name: Tilstone, William J. Main title: Fisher's techniques of crime scene investigation / William J. Tilstone, Michael L. Hastrup, Camilla Hald. Edition: 1st International Ed. Published/Created: Boca Raton, FL: CRC Press, c2013. Description: xxiv, 477 p.: ill. (chiefly col.); 27 cm. ISBN: 9781439817049 (hardcover: alk. paper) LC classification: HV8073.T556 2013 Related names: Hastrup, Michael L. Hald, Camilla. Contents: Prologue -- About the authors -- The forensic process, principles, and pract ice -- Forensic science and evidence -- The investigative star in crime scene investigation -- The forensic process: performance, practice, and procedures of scene investigation -- Practical scene investigation: the body in the woods -- Practitioner competency, professionalism, and codes of conduct -- Forms of evidence, identification, and recovery -- Establishing personal identity -- Trace evidence and miscellaneous materials -- Forensic biology -- Impression evidence -- Firearms examination -- Arson and explosions -- Sc ene investigation and type of crime -- Illicit drugs and toxicology -- Investigating sexual assault -- Burglary investigation -- Motor vehicle investigation -- Homicide investigation -- Appendix -- Bibliography -- Index. Subjects: Crime scene searches. Criminal investigation. Notes: Includes bibliographical references and index. Dewey class no.: 363.25

Forensic accounting and fraud examination LCCN: 2011001151 Type of material: Book Personal name: Hopwood, William S. Main title: Forensic accounting and fraud examination / William S. Hopwood, Jay J. Leiner, George R. Young. Edition: 2nd ed. Published/Created: New York: McGraw-Hill, c2012. Description: xv, 656 p.: ill.; 27 cm. ISBN: 9780078136665 (alk. paper) LC classification: KF8968.15.H67 2012 Related names: Leiner, Jay J. Young, George R. Contents: The forensic accounting legal environment -- Fundamentals I: accounting information systems -- Fundamentals II: the auditing environment -- Fraud prevention and risk management -- Fraud detection -- The fraud investigation and engagement processes -- The evidence collection process -- Fraud examination evidence I: physical, documentary, and observational evidence -- Fraud examination evidence II: interview and interrogation methods -- Fraud examination evidence III: forensic science and computer forensics -- The fraud report, litigation, and the recovery process -- Employee, vendor, and other frauds against the organization -- Financial statement fraud -- Fraud and SOX compliance -- Tax fraud -- Bankruptcy, divorce, and identity theft -- Organized crime, counterterrorism, and anti-money laundering -- Business valuation -- Dispute resolution services. Subjects: Forensic accounting--United States. Forensic accounting. Notes: Includes index. Dewey class no.: 363.25/6

Forensic biomechanics LCCN: 2012027882 Type of material: Book Personal name: Kieser, Jules. Main title: Forensic biomechanics / Jules Kieser, Michael Taylor, Debra Carr. Published/Created: Chichester, West Sussex, UK; Hoboken, NJ: Wiley-Blackwell, 2013. Description: viii, 164 p.: ill.; 25 cm. ISBN: 9781119990116 (hardback) LC classification:

QH513.K54 2013 Related names: Taylor, Michael, 1953 September 18- Carr, Debra D. Summary: "The book not only provides an in-depth examination of the field of trauma biomechanics, but also discusses the role of biomechanical principles in the analysis of blood-splatter patterning, assault and child abuse, ballistics and the investigation of the fracture of fibers and textiles"--Provided by publisher. Contents: Basic principles of biomechanics -- Biomechanics of bone and bony trauma -- Biomechanics of skin and soft tissue trauma -- The mechanics of bloodstain pattern formation -- Fibres and textiles. Subjects: Biomechanics. Forensic Sciences--methods. Wounds and Injuries--pathology. Notes: Includes bibliographical references and index. Series: Essentials of forensic science Dewey class no.: 617.1

Forensic chemistry handbook
LCCN: 2010053071 Type of material: Book Main title: Forensic chemistry handbook / edited by Lawrence Kobilinsky. Published/Created: Hoboken, N.J.: John Wiley & Sons, c2012. Description: xxii, 504 p., [16] p. of plates: ill. (some col.); 25 cm. ISBN: 9780471739548 (cloth) 0471739545 (cloth) 9781118062227 (ePDF) 1118062221 (ePDF) 9781118062241 (oBook) 1118062248 (oBook) 9781118062234 (ePub) 111806223X (ePub) LC classification: HV8073.F5595 2012 Related names: Kobilinsky, Lawrence F. edt Summary: The Forensic Chemistry Handbook focuses on topics in each of the major chemistry-related areas of forensic science. With chapter authors that span the forensic chemistry field, this book exposes readers to the state of the art on subjects such as serology (including blood, semen, and saliva), DNA/molecular biology, explosives and ballistics, toxicology, pharmacology, instrumental analysis, arson investigation, and various other types of chemical residue analysis. In addition, the Forensic Chemistry Handbook: covers forensic chemistry in a clear, concise, and authoritative way; brings together in one volume the key topics in forensics where chemistry plays an important role, such as blood analysis, drug analysis, urine analysis, and DNA analysis; explains how to use analytical instruments to analyze crime

scene evidence; contains numerous charts, illustrations, graphs, and tables to give quick access to pertinent information. Media focus on high-profile trials like those of Scott Peterson or Kobe Bryant have peaked a growing interest in the fascinating subject of forensic chemistry. For those readers who want to understand the mechanisms of reactions used in laboratories to piece together crime scenes--and to fully grasp the chemistry behind it--this book is a must-have. --Publisher description. Contents: 1. Forensic environmental chemistry / Anthony Carpi and Andrew J. Schweighardt -- 2. Principles and issues in forensic analysis of explosives / Jimmie C. Oxley, Maurice Marshall, and Sarah L. Lancaster -- 3. Analysis of fire debris / John J. Lentini -- 4. Forensic examination of soils / Raymond C. Murray -- 5. Analysis of paint evidence / Scott G. Ryland and Edward M. Suzuki -- 6. Analysis techniques used for the forensic examination of writing and printing inks / Gerald M. LaPorte and Joseph C. Stephens -- 7. The role of vibrational spectroscopy in forensic chemistry / Ali Koçak -- 8. Forensic serology / Richard Li -- 9. Forensic DNA analysis / Henrietta Margolis Nunno -- 10. Current and future uses of DNA microarrays in forensic science / Nathan H. Lents -- 11. Date-rape drugs with emphasis on GHB / Stanley M. Parsons -- 12. Forensic and clinical issues in alcohol analysis / Richard Stripp -- 13. Fundamental issues of postmortem toxicology / Donald B. Hoffman, Beth E. Zedeck, and Morris S. Zedeck -- 14. Entomotoxicology: drugs, toxins, and insects / Jason H. Byrd and Michelle R. Peace. Subjects: Chemistry, Forensic--Handbooks, manuals, etc. Forensic sciences--Handbooks, manuals, etc. Criminal investigation--Handbooks, manuals, etc. Notes: Includes bibliographical references and index. Dewey class no.: 363.25/62

Forensic DNA analysis: current practices and emerging technologies LCCN: 2013025792 Type of material: Book Main title: Forensic DNA analysis: current practices and emerging technologies / editors, Jaiprakash G. Shewale, Ph.D., technical director, Life Technologies, Foster City, California, USA; coeditor, Ray H. Liu, professor Emeritus, University of Alabama at Birmingham and editor-in-chief,

Forensic Science Review, Vancouver, Washington, USA. Published/Produced: Boca Raton: CRC Press, Taylor & Francis Group, [2014] Description: xvi, 405 pages: illustrations (some color); 26 cm ISBN: 9781466571266 (hardcover: alk. paper) LC classification: RA1057.F67 2014 Related names: Shewale, Jaiprakash G., editor of compilation. Liu, Ray H., coeditor of compilation. Contents: Forensic DNA evidence collection at a crime scene: an investigator's commentary / Joseph Blozis -- Optimizing storage and handling of DNA extracts / Steven B. Lee, Cecelia A. Crouse, and Margaret C. Kline -- Extraction of DNA from forensic biological samples for genotyping / James E. Stray, Jason Y. Liu, Maxim G. Brevnov, and Jaiprakash G. Shewale -- Extraction of DNA from human remains / James E. Stray and Jaiprakash G. Shewale -- RNA profiling for the identification of the tissue origin of dried stains in forensic biology / Erin K. Hanson and Jack Ballantyne -- Assessment of DNA extracted from forensic samples prior to genotyping / Maura Barbisin and Jaiprakash G. Shewale -- Principles, practice, and evolution of capillary electrophoresis as a tool for forensic DNA analysis / Jaiprakash G. Shewale, Liwei Qi, and Lisa M. Calandro -- Next-generation STR genotyping kits for forensic applications / Julio J. Mulero and Lori K. Hennessy -- Biology and genetics of new autosomal STR loci useful for forensic DNA analysis / John M. Butler and Carolyn R. Hill -- Hidden variation in microsatellite loci: utility and implications for forensic / John V. Planz and Thomas. A. Hall -- Additional Y-STRs in forensics: why, which, and when / Kaye N. Ballantyne and Manfred Kayser -- Forensic mitochondrial DNA analysis: current practice and future potential / Mitchell Holland, Terry Melton, and Charity Holland -- Applications of autosomal SNPs and indels in forensic analysis / Christopher Phillips -- Deep-sequencing technologies and potential applications in forensic DNA testing / Roxanne R. Zascavage, Shantanu J. Shewale, and John V. Planz -- Sample-to-result STR genotyping systems: potential and status / Jenny A. Lounsbury, Joan M. Bienvenue, and James P. Landers -- Training of forensic DNA scientists: a commentary / Meredith A. Turnbough, Arthur

J. Eisenberg, Lisa Schade, and Jaiprakesh G. Shewale. Subjects: Forensic Genetics--methods. DNA--analysis. Genotyping Techniques--methods. Notes: Includes bibliographical references and index. Dewey class no.: 614/.12

Forensic DNA applications: an interdisciplinary perspective LCCN: 2014395412 Type of material: Book Main title: Forensic DNA applications: an interdisciplinary perspective / edited by Dragan Primorac, M.D., Ph.D., Adjunct Professor of Forensic Science, Eberly College of Science, Pennsylvania State University, University Park, PA, USA, and Henry C. Lee College of Criminal Justice and Forensic Sciences, University of New Haven, West Haven, CT, USA, Professor of Pediatric Medicine, School of Medicine, University of Split, Split, Croatia and School of Medicine, University of Osijek, Osijek, Croatia; Moses Schanfield, Ph.D., Professor of Science and Anthropology, Department of Forensic Science, George Washington University, Mount Vernon Campus, Washington, D.C., USA. Published/Produced: Boca Raton: CRC Press, Taylor & Francis Group, [2014] ©2014 Description: xxv, 621 pages: illustrations (chiefly color); 26 cm ISBN: 9781466580220 (hardback: alk. paper) 1466580224 (hardback: alk. paper) LC classification: RA1057.5.F658 2014 Related names: Primorac, Dragan, editor. Schanfield, Moses S., editor. Subjects: Forensic genetics. DNA fingerprinting. DNA--Analysis. Forensic Genetics--methods. DNA--analysis. DNA Fingerprinting. Databases, Nucleic Acid--legislation & jurisprudence. Forensic Anthropology--methods. Notes: Includes bibliographical references and indexes. Dewey class no.: 614/.1

Forensic fraud: evaluating law enforcement and forensic science cultures in the context of examiner misconduct LCCN: 2012537599 Type of material: Book Personal name: Turvey, Brent E. author. Main title: Forensic fraud: evaluating law enforcement and forensic science cultures in the context of examiner misconduct / Brent E. Turvey. Published/Produced: Amsterdam: Elsevier, 2013. ©2013 Description: xi, 284 pages: color illustrations; 25 cm. ISBN: 9780124080737 0124080731 LC classification: HV8079.F7 T87 2013

Summary: Forensic Fraud is a timely work that examines the problem of forensic fraud both theoretically and empirically, assessing the relationships among examiner, workplace, evidentiary, and impact variables. Unvalidated or improper forensic science is a leading cause of wrongful convictions. In more than 50% of the DNA exonerations nationwide, unvalidated or improper forensic science contributed to the underlying wrongful conviction. These problems include forensic techniques that have not been subjected to rigorous scientific evaluation (such as hair microscopy, bite mark comparisons, firearm tool mark analysis and shoe print comparisons); testing that is improperly conducted or analysis that is not accurate (regardless of whether the forensic technique involved is validated); and forensic misconduct (such as fabricated test results and misleading testimony). Forensic Fraud is written by a forensic scientist for the forensic science community, rather than by legal scholars or practitioners, as has been the trend. Responds directly to the dramatic increase in forensic fraud related laboratory scandals and closures that have plagued the forensic science community. Offers realistic recommendations and reforms ideally suited for forensic science professionals, legal practitioners, laboratory supervisors, forensic science students and academicians. Subjects: Fraud investigation--United States. Criminal investigation--United States. Notes: Includes bibliographical references (pages 255-276) and index. Includes bibliographical references and index.

Forensic investigation of explosions LCCN: 2011031362 Type of material: Book Main title: Forensic investigation of explosions / edited by Alexander Beveridge. Edition: 2nd ed. Published/Created: Boca Raton, FL: CRC Press, c2012. Description: xxxvii, 792 p.: ill. (chiefly col.); 27 cm. Links: Contributor biographical information http://www.loc.gov/catdir/enhancements/fy1205/2011031362-b.html ISBN: 9781420087253 1420087258 LC classification: HV8079.B62 F67 2012 Related names: Beveridge, Alexander. Summary: "Now in its second edition, Forensic Investigation of Explosions draws on the editor's 30 years of explosives casework experience, including his work on task forces set up to investigate major explosives incidents. Dr. Alexander Beveridge provides a broad, multidisciplinary approach, assembling the contributions of internationally recognized experts who present the definitive reference work on the subject.Topics discussed include:The physics and chemistry of explosives and explosionsThe detection of hidden explosivesThe effect of explosions on structures and personsAircraft sabotage investigationsExplosion scene investigationsCasework managementThe role of forensic scientistsAnalysis of explosives and their residuesForensic pathology as it relates to explosivesPresentation of expert testimonyWith nearly 40 percent more material, this new edition contains revised chapters and several new topics, including:A profile of casework management in the UK Forensic Explosives Laboratory, one of the world's top labs, with a discussion of their management system, training procedures, and practical approaches to problem solvingProperties and analysis of improvised explosivesAn examination of the Bali bombings and the use of mobile analytical techniques and mobile laboratoriesThe collection, analysis, and presentation of evidence in vehicle-borne improvised explosive device cases, as evidenced in attacks on US overseas targetsThis volume offers valuable information to all members of prevention and post-blast teams. Each chapter was written by an expert or experts in a specific field and provides well-referenced information underlying best practices that can be used in the field, laboratory, conference room, classroom, or courtroom. "--Provided by publisher. Subjects: Bombing investigation. Explosives--Detection. Explosions. Chemistry, Forensic. Notes: Includes bibliographical references and index. Series: International forensic science and investigation series International forensic science and investigation series. Dewey class no.: 363.325/165

Forensic investigation of unusual firearms: ballistic and medico-legal evidence LCCN: 2014021769 Type of material: Book Personal name: Sinha, J. K. Main title: Forensic investigation of unusual firearms: ballistic and medico-legal evidence / J.K. Sinha, PhD, former Director In-Charge and Deputy Director (Ballistics), Central Forensic Science Laboratory Chandigarh, India. Published/Produced: Boca Raton: CRC, Taylor & Francis Group, [2015] Description: xxi, 190 pages; 25 cm ISBN: 9781466571372 LC classification: HV8077.S57 2015 Subjects: Forensic ballistics. Firearms--Identification. Firearms--Markings. Notes: "A CRC title." Includes bibliographical references (pages 175-177) and index. Dewey class no.: 363.25/62

Forensic metrology: scientific measurement and inference for lawyers, judges and criminalists LCCN: 2014024204 Type of material: Book Personal name: Vosk, Ted, author. Main title: Forensic metrology: scientific measurement and inference for lawyers, judges and criminalists / Ted Vosk, Cowan Kirk Gaston Wolff Law Firm, Kirkland, Washington, USA; Ashley F. Emery, University of Washington, Seattle, Washington, USA. Published/Produced: Boca Raton: CRC Press,Taylor & Francis Group, [2015] Description: xlvii, 448 pages; 24 cm. ISBN: 9781439826195 (hardback: acid-free paper) LC classification: HV8073.5.V67 2015 Related names: Emery, A. F. (Ashley Francis), 1934- Summary: "With contributions from professionals in forensic science, law, and engineering, this book serves as an introduction to the field of metrology, its application to the forensic sciences, and its use in the courtroom. Written for professionals and students, the text begins at a general level, demonstrating that the principles of metrology are familiar to all. It then builds a more sophisticated level of coverage, combining the scientific machinery of metrology and forensics with the practice of law so practitioners will be able to apply the principles discussed. Examples, case studies, and diagrams are included for ease of understanding and application"-- Provided by publisher. "Foreword Facts are inherently nonexplanatory. A report of "70 degrees" means nothing without some context: Is it January or July? Juneau or Jakarta? Celsius or Fahrenheit? Shewhart's dictum ("Data has no meaning apart from its context") is central to all sciences, including--perhaps particularly--forensic science. A few assumptions underly this concept. First is that the context trumps the data in that, as Deming said, "Experience without theory teaches nothing. In fact, experience can not even be recorded unless there is some theory, however crude... " (Deming, 1986, p. 317). If you must be "this tall" to ride this roller coaster, then in the context (the roller coaster's safety design), height (the measurement) is important because it is ultimately based on theory (human biological height is predictably proportional to age and weight): Children of a certain age are large enough to be protected by the coaster's safety mechanisms, which were designed with bodies of a minimal size in mind. Simply saying a child is 40 inches tall means very little. The second assumption is that there are data, that is, plural. Science is based on reproducibility and with that comes the collection of multiple data points, either by ourselves to ensure accuracy or by others to check validity. In each measurement is a minor bit of error (in the statistical sense) and multiple measurements help us capture, understand, and control that error. Twenty one different Six Flags Amusement Parks exist and each one has different roller coasters, ranging from children's coasters to Mega Coasters, all with different height requirements (Table 0.1). Multiple measurements are taken, recorded, and communicated to each of Six Flags' parks to ensure that the requirements are consistent between parks"-- Provided by publisher. Subjects: Forensic sciences. Metrology. Evidence (Law) LAW / Criminal Law / General. LAW / Forensic Science. Notes: Includes bibliographical references and index. Series: International forensic science and investigation Dewey class no.: 363.25

Forensic science advances and their application in the judiciary system LCCN: 2011027734 Type of material: Book Main title: Forensic science advances and their application in the judiciary system / editors, Danielle Sapse, Lawrence Kobilinsky. Published/Created: Boca Raton: CRC Press, c2012. Description: xii, 283 p.: ill.; 27 cm. ISBN: 9781439829592 (hardcover: alk. paper) LC classification: HV8073.F5833 2012 Related names: Sapse, Danielle S. Kobilinsky, Lawrence F. Subjects: Forensic sciences. Criminal investigation. Notes: Includes bibliographical references and index. Dewey class no.: 363.25

Forensic science and the administration of justice: critical issues and directions LCCN: 2013046876 Type of material: Book Main title: Forensic science and the administration of justice: critical issues and directions / Kevin J. Strom, RTI International, Matthew J. Hickman, Seattle University. Published/Produced: Los Angeles: Sage, [2015] ©2015 Description: xv, 293 pages; 26 cm ISBN: 9781452276885 (pbk.) 1452276889 (pbk.) LC classification: HV8073.F5967 2015 Related names: Strom, Kevin, editor. Hickman, Matthew J., editor. Contents: Preface -- Acknowledgments -- The demand for forensic services -- A historical review of the demand for forensic evidence / Joseph L. Peterson -- Is there evidence of a "csi effect"? / Rachel Dioso-Villa -- What we know (and don't know) about evidence backlogs / Matthew J. Hickman and Kevin J. Strom -- The quality of forensic services -- Adopting a research culture in the forensic sciences / Barry A. J. Fisher -- Minimizing contextual bias in forensic casework / Reinoud Stoel, Charles Berger, Wim Kerkhoff, Erwin J. A. T. Mattijssen, and Itiel Dror -- A survey of ethical issues in the forensic sciences / Jay Siegel -- The utility of forensic services -- The impact of forensic evidence on criminal justice: evidence from case -- Processing studies / Sally Kelty, Roberta Julian, and Robert Hayes -- Assessing the utility of dna evidence in criminal investigations / Michael D. White, Andrea R. Borrego, and David A. Schroeder -- Forensic science: the prosecutor's role / Nina Chernoff -- Post-conviction issues -- The problems and challenges of evidence retention / John M. Collins, Jr -- Innovation, success, error, and confidence in forensic dna testing / Kristen Skogerboe -- Developing new business models for forensic laboratories / Max M. Houck and Paul J. Speaker -- Rethinking the role of the crime laboratory in criminal justice decision -- Making / Kevin J. Strom and Matthew J. Hickman -- The future of forensic science / Walter F. Rowe -- Index -- About the editors -- About the contributors. Subjects: Forensic sciences. Justice, Administration of. Notes: Includes bibliographical references and index. Dewey class no.: 363.25

Forensic science and the law: a guide for lawyers, police and expert witnesses LCCN: 2014427315 Type of material: Book Personal name: Sandiford, Anna, 1970- author. Main title: Forensic science and the law: a guide for lawyers, police and expert witnesses / Dr. Anna Sandiford. Published/Produced: Wellington: Thomson Reuters, [2013] ©2013 Description: xvii, 385 pages: illustrations; 24 cm. Links: Table of contents only http://www.loc.gov/catdir/toc/fy14pdf05/2014427315.html ISBN: 9780864728418 0864728417 LC classification: KUQ4700.S26 2013 Subjects: Forensic sciences--New Zealand. Evidence, Expert--New Zealand. Evidence (Law)--New Zealand. Evidence (Law) Forensic sciences. New Zealand. Notes: Cover subtitle: A guide for police, lawyers and expert witnesses. Includes bibliographical references and index. Series: Silver series Silver series (Wellington, N.Z.) Dewey class no.: 363.250993

Forensic science evidence: can the law keep up with science? LCCN: 2012012955 Type of material: Book Personal name: Shelton, Donald E. Main title: Forensic science evidence: can the law keep up with science? / Donald E. Shelton. Published/Created: El Paso: LFB Scholarly Pub. LLC, 2012. Description: vii, 273 p.; 23 cm. ISBN: 9781593325176 (hardcover: alk. paper) 1593325177 (hardcover: alk. paper) LC classification: KF9674.S529 2012 Subjects: Evidence, Expert--United States. Forensic sciences--United States. Evidence, Criminal--United States. Notes: Includes bibliographical references (p. 227-270) and index. Series: Criminal justice: recent scholarship Dewey class no.: 345.73/067

Forensic science in contemporary American popular culture: gender, crime, and science LCCN: 2012020140 Type of material: Book Personal name: Steenberg, Lindsay, 1976- Main title: Forensic science in contemporary American popular culture: gender, crime, and science / by Lindsay Steenberg. Published/Created: New York: Routledge, c2013. Description: xii, 221 p.: ill.; 24 cm. ISBN: 9780415891882 LC classification: HV6789.S642 2012 Contents: Retrofit forensics: excavating the mythic origins of forensic science -- The forensic women's gothic: the postfeminism and expertise --

Forensic archaeologies on crime television: science and gender as seen on TV -- The CSI effect -- Forensic science at the museum -- Forensic science and lifestyle -- Conclusion: the evidence lies. Subjects: Crime in popular culture--United States. Forensic sciences--Social aspects--United States. Criminal investigation in mass media. Notes: Includes bibliographical references and index. Series: Routledge research in cultural and media studies; 45 Dewey class no.: 363.250973

Forensic science in criminal investigation & trials LCCN: 2014360417 Type of material: Book Personal name: Sharma, B. R. (Forensic expert) Main title: Forensic science in criminal investigation & trials / Dr. B.R. Sharma, M.Sc. (Hons.), Ph.D., F.A.F.Sc. Edition: Fifth edition. Published/Produced: New Delhi, India: Universal Law Publishing Co. Pvt. Ltd., 2014. Description: lxxxvii, 1732 pages: illustrations (black and white); 25 cm ISBN: 9789350354681 9350354683 LC classification: LAW+ KNS4636.S53 2014 Subjects: Criminal investigation--India. Evidence, Expert--India. Forensic sciences--India. Notes: Includes bibliographical references and index. Reproduction no./Source: Library of Congress -- New Delhi Overseas Office Dewey class no.: 345.54052

Forensic science LCCN: 2012010306 Type of material: Book Personal name: Lyle, D. P. Main title: Forensic science / D. P. Lyle. Edition: 1st ed. Published/Created: Chicago: American Bar Association, [Criminal Justice Section], c2012. Description: xviii, 420 p.: ill.; 23 cm. ISBN: 9781614383529 (pbk.: alk. paper) LC classification: KF8961.L95 2012 Related names: American Bar Association. Section of Criminal Justice Contents: Basic considerations and history -- General organization of forensic science -- Evidence -- The autopsy -- Corpse identification -- Time of death -- Cause, mechanism, and manner of death -- Traumatic injuries -- Asphyxia -- Blood and other bodily fluids -- DNA -- Toxicology -- Fingerprints -- Bloodstains -- Impressions: shoes, tires, and tools -- Trace evidence -- Firearms examination -- Arson investigation -- Questioned documents -- Criminal psychology. Subjects: Evidence,

Expert--United States. Forensic sciences--United States. Notes: Includes index. Dewey class no.: 363.250973

Forensic science: an introduction to scientific and investigative techniques LCCN: 2013025443 Type of material: Book Main title: Forensic science: an introduction to scientific and investigative techniques / edited by Stuart H. James, Jon J. Nordby, Suzanne Bell. Edition: Fourth edition. Published/Produced: Boca Raton, Florida: CRC Press/Taylor & Francis Group, [2014] Description: xiv, 600 pages: color illustrations; 29 cm ISBN: 9781439853832 (hbk.: alk. paper) LC classification: HV8073.F5835 2014 Related names: James, Stuart H., editor of compilation. Nordby, Jon J., editor of compilation. Bell, Suzanne., editor of compilation. Subjects: Forensic sciences--Handbooks, manuals, etc. Criminal investigation--Handbooks, manuals, etc. Evidence, Criminal--Handbooks, manuals, etc. Criminal laboratories--Handbooks, manuals, etc. Notes: Includes bibliographical references and index. Dewey class no.: 363.25

Forensic science: current issues, future directions LCCN: 2012026190 Type of material: Book Main title: Forensic science: current issues, future directions / Douglas H. Ubelaker, editor. Published/Created: Chichester, West Sussex; Hoboken, NJ: Wiley-Blackwell, 2013. Description: xviii, 404 p.: ill.; 26 cm. ISBN: 9781119941231 (cloth) LC classification: HV8073.F5836 2013 Related names: Ubelaker, Douglas H. Subjects: Forensic sciences. Notes: Includes bibliographical references and index. Dewey class no.: 363.25

Forensic science: from the crime scene to the crime lab LCCN: 2011038163 Type of material: Book Personal name: Saferstein, Richard, 1941- Main title: Forensic science: from the crime scene to the crime lab / Richard Saferstein. Edition: 2nd ed. Published/Created: Boston: Pearson/Prentice Hall, c2013. Description: xxi, 546 p.: ill. (chiefly col.); 28 cm. ISBN: 9780131391871 0131391879 LC classification: HV8073.S214 2013 Contents: Introduction -- Definition and scope of forensic science -- Securing and searching the crime scene -- Recording the crime scene --

Collection of crime-scene evidence -- Physical evidence -- Death investigation -- Crime-scene reconstruction -- Fingerprints -- Firearms, toolmarks, and other impressions -- Bloodstain pattern analysis -- Drugs -- Forensic toxicology -- Trace evidence I: hairs and fibers -- Trace evidence II: paint, glass, and soil -- Biological stain analysis: DNA -- Forensic aspects of fire and explosion investigation -- Document examination -- Computer forensics -- Appendix I: guides to the collection of physical evidence: FBI -- Appendix II: instructions for collecting gunshot residue -- Appendix III: chemical formulas for latent fingerprint development -- Appendix IV: Chemical formulas for development of footwear impressions in blood -- Answers to end-of-chapter questions -- Index. Subjects: Criminal investigation. Crime scene searches. Forensic sciences. Evidence, Criminal. Notes: Includes index. Dewey class no.: 363.25

Forensic science: in pursuit of justice LCCN: 2014943866 Type of material: Book Personal name: Carmichael, L. E. (Lindsey E.), author. Main title: Forensic science: in pursuit of justice / by L.E. Carmichael. Published/Created: North Mankato, MN: Essential Library, c2015. Description: 112 pages: color illustrations; 22 x 25 cm. ISBN: 1624035612 9781624035616 Summary: "This title presents the history of forensics. Vivid text details how early studies of toxic chemicals and firearm analysis led to modern scientific crime solving techniques. It also puts a spotlight on the brilliant scientists who made these advances possible."--Publisher's website. Subjects: Forensic sciences--History--Juvenile literature. Criminal investigation--Juvenile literature. Notes: Includes bibliographical references and indexes. Series: History of science History of science (Essential Library (Firm)) Dewey class no.: 363.25

Forensic Testimony: Science, Law and Expert Evidence LCCN: 2013417483 Type of material: Book Personal name: Bowers, C. Michael, author. Main title: Forensic Testimony: Science, Law and Expert Evidence / C. Michael Bowers. Published/Produced: Kidlington, Oxford; San Diego, CA: Academic Press, 2014.

Description: xxxii, 263 pages: illustrations; 25 cm ISBN: 9780123970053 0123970059 LC classification: KF8961.B69 2014 Contents: 1. The history of experts in English common law, with practice advice for beginning experts / C. Michael Bowers -- 2. Science and forensic science / Marc Page -- 3. The admissibility of forensic expert evidence / Mark Page -- 4. Professional forensic expert practice / Mark Page -- 5. Managing your forensic case from beginning to end: it's all about communication / C.M. Bowers -- 6. Character traits of expert witnesses: the good and the bad / C.M. Bowers -- 7. Voir dire and direct examination of the expert / C.M. Bowers -- 8. Cross examination: the expert's challenge and the lawyer's strategies / C.M. Bowers -- 9. Uniqueness and individualization in forensic science / Mark Page -- 10. Forensic failures / Brent E. Turvey -- 11. Forensic expert ethics / C.M. Bowers -- 12. The unparalleled power of expert testimony / Wendy J. Koen. Subjects: Evidence, Expert--United States. Evidence, Expert. Forensic sciences. Notes: Includes bibliographical references and index.

Foundations of forensic document analysis: theory and practice LCCN: 2015006398 Type of material: Book Personal name: Allen, Michael (Michael John), 1959-, author. Main title: Foundations of forensic document analysis: theory and practice / Michael Allen. Published/Produced: Chichester, West Sussex; Hoboken, NJ: John Wiley & Sons, Ltd, 2015. Projected pub date: 1505 Description: p.; cm. ISBN: 9781118646892 (cloth) LC classification: RA1147.5 Contents: The comparison of handwriting -- Signature examination -- The examination of typed and copy documents -- The examination of printed documents -- Materials used to create documents -- Analytical techniques used in document examination -- Altered documents -- Indented impressions -- Dating documents -- Duties of the expert Subjects: Documentation. Forensic Sciences--methods. Records as Topic. Notes: Includes bibliographical references and index. Additional formats: Online version: Allen, Michael (Michael John), 1959-, author. Foundations of forensic document analysis Chichester, West Sussex; Hoboken, NJ: John Wiley & Sons, Ltd, 2015

9781118729915 (DLC) 2015007607 Series: Essentials of forensic science Essentials of forensic science (Forensic Science Society) Dewey class no.: 614/.1

Fundamentals of forensic science LCCN: 2015937705 Type of material: Book Personal name: Houck, Max. Main title: Fundamentals of forensic science / Max Houck, Jay Siegel. Published/Produced: Boston, MA: Elsevier, 2015. Projected pub date: 1510 Description: pages cm ISBN: 9780128000373.

Illustrated guide to home forensic science experiments: all lab, no lecture LCCN: 2012406837 Type of material: Book Personal name: Thompson, Robert Bruce. Main title: Illustrated guide to home forensic science experiments: all lab, no lecture / Robert Bruce Thompson and Barbara Fritchman Thompson. Edition: 1st ed. Published/Created: Farnham: O'Reilly, c2012. Description: xviii, 425 p.: col. ill.; 25 cm. ISBN: 9781449334512 (pbk.) 1449334512 (pbk.) LC classification: HV8073.8.T54 2012 Related names: Thompson, Barbara Fritchman. Subjects: Forensic sciences--Popular works. Forensic sciences--Experiments--Popular works. Notes: Includes index. Series: DIY science. DIY science. Dewey class no.: 363.25

Improving forensic science in the criminal justice system: hearing before the Committee on the Judiciary, United States Senate, One Hundred Twelfth Congress, second session, July 18, 2012. LCCN: 2012518177 Type of material: Book Corporate name: United States. Congress. Senate. Committee on the Judiciary. Main title: Improving forensic science in the criminal justice system: hearing before the Committee on the Judiciary, United States Senate, One Hundred Twelfth Congress, second session, July 18, 2012. Published/Created: Washington: U.S. G.P.O.: For sale by the Supt. of Docs., U.S. G.P.O., 2012. Description: iii, 125 p.: ill.; 24 cm. Links: PDF version: http://purl.fdlp.gov/GPO/gpo28789 LC classification: KF26.J8 2012a Subjects: Forensic sciences--United States--Evaluation. Forensic sciences--Standards--United States. Evidence, Criminal--Standards--United States. Judicial error--United States--Prevention. Notes: Shipping list no.: 2013-0054-P. "Serial no. J-112-86."

Includes bibliographical references. Additional formats: Also available via the Internet from the GPO Access web site. Address as of 01/11/13: http://purl.fdlp.gov/GPO/gpo28789; current access is available via PURL. Online version: United States. Congress. Senate. Committee on the Judiciary. Improving forensic science in the criminal justice system (OCoLC)811247293 Series: S. hrg.; 112-519 United States. Congress. Senate. S. hrg.; 112-519. Government doc no.: Y 4.J 89/2:S.HRG.112-519 GPO item no.: 1042-A Report no.: Serial no. J-112-86 (United States. Congress. Senate. Committee on the Judiciary)

In-depth Creation basics & beyond: an in-depth look at science, origins, and evolution LCCN: 2013944862 Type of material: Book Main title: In-depth Creation basics & beyond: an in-depth look at science, origins, and evolution / Henry Morris III, John D. Morris, Jason Lisle, James J.S. Johnson, Nathaniel Jeanson, Randy Guliuzza, Jeffrey Tomkins, Jake Hebert, Frank Sherwin, Brian Thomas. Published/Produced: Dallas, Tex.: Institute for Creation Research, 2013. Description: 348 pages: illustrations, maps; 23 cm ISBN: 9781935587309 (pbk.) 1935587307 (pbk.) LC classification: BS651.I53 2013 Variant title: In-depth Creation basics and beyond Creation basics and beyond Spine title: Creation basics & beyond Related names: Morris, Henry M. Contents: Introduction / Jason Lisle -- WORLDVIEW: GOD DETERMINES TRUTH OR MAN DETERMINES TRUTH? -- 1. Genesis and the character of God / Henry Morris III -- 2. Genesis is relevant to Christian doctrines / Jason Lisle -- 3. Day-age theory: a day late and a scholar short / Jason Lisle and James J. S. Johnson -- 4. Gap theory: a formless and void error / Jason Lisle and James J. S. Johnson -- 5. Genesis is narrative history, not Hebrew poetry / James J. S. Johnson -- 6. Genesis data add up to a young Earth / James J. S. Johnson -- 7. No death before Adam sinned / James J. S. Johnson -- 8. Evolution is logically impossible / Jason Lisle -- 9. How should we then interpret Genesis? / Jason Lisle and James J. S. Johnson -- 10. Fossil forensics disprove Darwin / James J. S. Johnson -- 11. Evolutionists flunk basic forensic science / James J. S. Johnson -- 12. Evolutionist vocabulary on cellular communication displays faulty logic / James J. S. Johnson -- 13. Only biblical creation proves God loves you personally / James J. S. Johnson -- BIOLOGY: CREATED KINDS OR COMMON ANCESTRY? -- 14. All-or-nothing unity in the human body / Randy J. Guliuzza -- 15. The mistakes in evolutionary arguments against life's design / Randy Guliuzza -- 16. Unmistakable evidence for God's design: cells lead the way / Nathaniel Jeanson and Brian Thomas -- 17. The origin of species: did Darwin get it right? / Nathaniel Jeanson -- 18. Does biological similarity prove evolutionary ancestry? / Nathaniel Jeanson -- 19. The erosion of genes confirms Genesis history / Brian Thomas -- 20. Is the evolutionary tree of life real? / Jeffrey Tomkins, Frank Sherwin and Brian Thomas -- 21. The junk DNA scam / Jeffrey Tomkins -- 22. The origin of predator/prey relationship and parasites / Frank Sherwin -- 23. Cain, his wife, and the origin of races / Frank Sherwin -- 24. Ape-man or image of God? / Jeffrey Tomkins -- 25. Chromosome 2: fusion or not? / Jeffrey Tomkins -- 26. The mystery of life's beginning / Jeffrey Tomkins -- GEOLOGY: RECENT FLOOD OR MILLIONS OF YEARS? -- 27. The Flood is the key / John D. Morris -- 28. The Flood was global according to Scripture / John D. Morris -- 29. The Flood was global and catastrophic according to geology / John D. Morris -- 30. The young Earth in Scripture / John D. Morris -- 31. The young Earth in geology / John D. Morris -- 32. The real nature of the fossil record / John D. Morris -- 33. Do radioisotope dating methods prove an old Earth? / Jake Hebert -- 34. Was there an Ice Age? / Jake Hebert -- 35. Do seafloor sediment and ice core data prove long ages? / Jake Hebert -- 36. Catastrophic plate tectonics / John D. Morris -- 37. Developing a sensible approach to human origins research, part 1: Fossils and bias / Randy J. Guliuzza -- 38. Developing a sensible approach to human origins research, part 2: Understanding the fossils / Randy J. Guliuzza -- DINOSAURS AND MAN: WALKED TOGETHER OR RULED APART? -- 39. Dinosaurs and the Bible / Frank Sherwin -- 40. Original tissue fossils cannot be millions of years old / Brian Thomas -- 41. Were there really feathered dinosaurs? / Brian Thomas -- 42. How do they date a fossil anyway? / Brian Thomas -- 43.

Dinosaurs and dragon legends / Brian Thomas -- ASTRONOMY: CREATED COSMOS OR THE BIG BANG? -- 44. A big bang, really? / Jason Lisle -- 45. The age of the cosmos: what you have not been told / Jason Lisle -- 46. Could the laws of physics have created the universe? / Jake Hebert -- 47. Distant starlight in a young universe / Jason Lisle -- 48. Do UFOs exist? / Jason Lisle -- 49. Do other universes exist? / Jake Hebert -- Conclusion / Jason Lisle -- Appendix: The scientific case against evolution / Henry M. Morris. Subjects: Creationism. Evolution. Bible and science. Notes: Includes bibliographical references and index.

Infrared and Raman spectroscopy in forensic science LCCN: 2011037212 Type of material: Book Main title: Infrared and Raman spectroscopy in forensic science / [edited by] John M. Chalmers, Howell G. M. Edwards, Michael D. Hargreaves. Edition: 1 Published/Created: Chichester, West Sussex, UK; Hoboken: Wiley, 2012. Projected pub date: 1204 Description: p. cm. Links: Cover image http://catalogimages.wiley.com/images/db/jimages/9780470749067.jpg ISBN: 9780470749067 (hardback) LC classification: HV8073.I4424 2012 Related names: Chalmers, John M. Edwards, Howell G. M., 1943- Hargreaves, Michael D. Summary: "This book will provide a survey of the major areas in which information derived from vibrational spectroscopy investigations and studies have contributed to the benefit of forensic science, either in a complementary or a unique way. This is highlighted by examples taken from real case studies and analyses of forensic relevance, which provide a focus for current and future applications and developments"-- Provided by publisher. Contents: Machine generated contents note: About the Editors xixList of Contributors xxiPreface xxvSECTION I: INTRODUCTION 11 Introduction and scope 31.1 Historical prologue 31.2 The application of infrared spectroscopy and Raman spectroscopy in forensic science 5References 62 Vibrational spectroscopy techniques: basics and instrumentation 72.1 Introduction 72.2 Vibrational spectroscopy techniques 72.2.1 The basics and some comparisons 72.2.1.1

Neither FT-Raman nor traditional dispersive Raman 2295.2.6 Surface enhanced Raman and surface enhanced resonance Raman studies of explosives 2295.2.7 Dispersive Raman studies of explosives 2305.2.8 Compact dispersive Raman spectrometers for the study of explosives 2315.2.9 Spatially offset Raman spectroscopy 2355.2.10 Stand-off Raman of explosives 2365.2.11 Raman microscopy and imaging 2365.2.12 Vehicle-mounted Raman analysers 2375.2.13 Classification schema for explosives 2375.2.14 Summary 238References 2385.3 Handheld Raman and FT-IR spectrometers 2435.3.1 Introduction 2435.3.2 Handheld/portable Raman and FT-IR devices 2435.3.3 Chemical warfare agents 2445.3.4 Explosives 2455.3.5 Tactical considerations 2465.3.6 Sample considerations 2465.3.7 Raman and FT-IR spectroscopy explosive identification capabilities 2485.3.8 Performance characterisation 2525.3.9 Summary 252References 2535.4 Non-invasive detection of concealed liquid and powder explosives using spatially offset Raman spectroscopy 2555.4.1 Introduction 2555.4.2 Discussion and examples 2565.4.3 Summary 258References 2595.5 Terahertz frequency spectroscopy and its potential for security applications 2615.5.1 Introduction 2615.5.2 Terahertz frequency radiation 2615.5.3 Terahertz time-domain spectroscopy 2625.5.4 Example Subjects: Forensic sciences. Infrared spectroscopy. Raman spectroscopy. Criminal investigation. SCIENCE / Spectroscopy & Spectrum Analysis Notes: Includes bibliographical references and index. Dewey class no.: 363.25/6

Introduction in law enforcement LCCN: 2012037623 Type of material: Book Main title: Introduction in law enforcement / David H. McElreath... [et al.]. Published/Created: Boca Raton, FL: CRC Press, c2013. Description: xix, 459 p.: ill.; 27 cm. ISBN: 9781466556232 (hbk.: alk. paper) LC classification: HV8139.I58 2013 Related names: McElreath, David. Contents: Crime in the United States -- The history of law enforcement in the United States -- Local and trubal enforcement in the United States -- State law enforcement in the United States -- Federal and international law enforcement -- Private enforcement -- Ethics -- Types of policing -- Patrol types and specialized assignmnents -- Investigation -- Forensic science -- After arrest: taking the case to court -- Law enforcement under the United States Constitution -- Administration and leadership: community support, recruitment, selection, training and retention -- The future of law enformcement and its changing role. Subjects: Law enforcement--United States. Police--United States. Notes: Includes bibliographical references and index. Dewey class no.: 363.20975

Introduction to criminal investigation LCCN: 2011034571 Type of material: Book Main title: Introduction to criminal investigation / edited by Michael L. Birzer and Cliff Roberson. Published/Created: Boca Raton, FL: CRC Press, c2012. Description: xvi, 384 p.: ill.; 26 cm. ISBN: 9781439839348 (pbk.) 1439839344 (pbk.) LC classification: HV8073.I568 2012 Related names: Birzer, Michael L., 1960- Roberson, Cliff, 1937- Contents: Foundations -- A brief history of criminal investigation / John A. Eterno -- Introducing criminal investigation / Bryan Courtney -- Preliminary considerations -- Crime scene search / Michael L. Birzer -- Report writing / Gene L. Scaramella -- Interview and interrogation / Scott M. Mire and Robert D. Hanser -- Evidence / Cory Rodivich -- Role of forensic science / Cory Rodivich -- Property crimes and financial investigations -- Vandalism / Matthew O'Deane -- Larceny and burglary / Walter J. Wywadis -- Auto theft / Donald Munday -- Arson investigations / Cliff Roberson -- Financial investigation / Michael J. Palmiotto -- Crimes against persons -- Homicide and ass ault / Gregg W. Etter and Roger L. Pennel -- Sex crimes / John Padgett -- Robb ery / J. Harrison Watts -- Specialized investigations -- Narcotics investigation / Donald F. Vespa -- Cybercrime / Mark R. McCoy -- Gang investigation / Gregg W. Etter -- Putting it all together -- Legal iss ues in criminal investigations / Frank DiMarin -- Preparing the case for court / Cliff Roberson and Gwynne Birzer -- Index. Subjects: Criminal investigation. Crime. Notes: Includes bibliographical references and index. Dewey class no.: 363.25

Introduction to policing LCCN: 2012037731 Type of material:

Book Personal name: Cox, Steven M. Main title: Introduction to policing / Steven M. Cox, William P. McCamey, Gene L. Scaramella. Edition: 2nd ed. Published/Created: Thousand Oaks: SAGE Publications, c2014. Description: xvii, 460 p.: col. ill.; 28 cm. ISBN: 9781452256610 (pbk.) 1452256616 (pbk.) LC classification: HV8139.S32 2014 Related names: McCamey, William. Scaramella, Gene L. Contents: Machine generated contents note: pt. I FOUNDATIONS OF POLICING -- ch. 1 Policing in America -- Chapter Learning Objectives -- Brief Overview of the Roles and Functions of Police in Society -- The Concept of Police -- Size and Scope of the Law Enforcement Sector -- Levels of Policing -- Additional Types of Police -- State Police -- Special Jurisdiction Police -- Sheriffs and Deputy Sheriffs -- Auxiliary/Reserve/Special Police -- Conservation Police Officers, Game Wardens -- Tribal Police Officers -- Chapter Summary -- Key Terms -- Discussion Questions -- Internet Exercises -- ch. 2 A Brief History of Police in the United States -- Chapter Learning Objectives -- Evolution of American Policing: An Overview -- Early History of Policing -- The Political Era -- The Reform Era -- Policing in the 1960s and 1970s -- The Community Policing/Problem Solving Era (1980-Early 21st Century) -- Some Contemporary Policing Strategies -- Areal Policing -- Reassurance Policing -- Intelligence-Led or Intelligence-Based Policing -- Terrorism-Oriented Policing -- Diversity in Police Services -- Current Issues in Policing -- Chapter Summary -- Key Terms -- Discussion Questions -- Internet Exercises -- ch. 3 Police Organization and Administration: An Overview -- Chapter Learning Objectives -- Organizational Structures -- Police Hierarchy and the Pyramid -- The Paramilitary Structure -- Functional Design -- Operations Division -- Patrol -- Investigations -- Administrative or Staff Services Division -- Place Design -- Time Design -- Future Organizational Designs -- Police Organizations in Context -- Change in Police Organizations -- Police Leadership -- Police Chiefs -- Police Unions, Professionalism, and Accreditation -- Police Unions and Collective Bargaining -- Police Professionalism -- Police Unions

Community Policing Efforts -- Changes and Innovations in Policing Strategies -- Intelligence-Led Policing -- Situational Crime Prevention -- Hot-Spot Policing -- Directed Patrol -- Differential Response Policing -- Saturation Patrol and Crackdowns -- Third-Party Policing -- Evidence-Based Policing -- Pulling Levers Policing -- Broken Windows Policing -- CompStat -- Incident Command Systems -- Dual Career Ladder -- Procedural Justice Policing -- Chapter Summary -- Key Terms -- Discussion Questions -- Internet Exercises -- pt. III POLICE CONDUCT -- ch. 8 The Police Culture -- Chapter Learning Objectives -- The Police Subculture -- Danger, Authority, and Efficiency -- The Police Personality: Myth or Reality? -- Stresses and Strains of Police Work -- The Personal Costs of Police Work -- Task Demands -- Role Demands -- Role Ambiguity -- Interpersonal Demands -- Physical Demands -- Forms of Police Stress -- Stages of Stress -- Burnout -- Stress and Police Families -- Police Shootings and Critical Incidents as a Source of Stress -- Attempts to Combat Police Stress -- Dealing With Stress in Police Organizations -- Chapter Summary -- Key Terms -- Discussion Questions -- Internet Exercises -- ch. 9 Law, Court Decisions, and the Police -- Chapter Learning Objectives -- The First Amendment -- The Second Amendment -- The Fourth Amendment -- Probable Cause and Reasonableness -- Searches and Seizures With and Without a Warrant -- Police Stops -- Police Searches Incident to Arrest -- Consent Searches -- The Fifth Amendment -- The Fourteenth Amendment -- The Exclusionary Rule -- Police Use of Force -- The Patriot Act, Homeland Security, and Terrorism -- Chapter Summary -- Key Terms -- Discussion Questions -- Internet Exercises -- ch. 10 Discretion, Ethics, and Accountability in Policing -- Chapter Learning Objectives -- Introduction -- Police Discretion -- Biased Enforcement and Racial Profiling -- Controlling Discretion and Encouraging Ethical Behavior -- Commitment to the Importance of Ethical Decision Making -- Improved Hiring Practices -- Elimination of Quotas -- Intolerance of Malfeasance -- In-Car Video Cameras -- Positive Media Relations -- Ethics in Policing -- Police Accountability -- Professionalism and the Police -- Chapter Summary -- Key Terms

-- Fingerprints -- DNA -- Police Use of Speed Detection Devices or Systems -- Body Armor and Weapons -- Police Weapons -- Chapter Summary -- Key Terms -- Discussion Questions -- Internet Exercises -- ch. 14 Terrorism, Homeland Security, and Global Issues -- Chapter Learning Objectives -- Introduction -- Transnational Crime -- Organized Crime -- White-Collar Crime -- Terrorism -- Homeland Security and the Police Response -- Department of Homeland Security -- The USA Patriot Act -- Racketeer Influenced and Corrupt Organizations (RICO) Statute -- First Responder Preparedness and the Role of the Public -- Chapter Summary -- Key Terms -- Discussion Questions -- Internet Exercises -- ch. 15 Private and Contract Police -- Chapter Learning Objectives -- History and Background -- The Development of the Modern Security Era -- Security Today -- Terrorism and Modern Private Security Operations -- Private and Contract Security Personnel -- Private Detectives and Investigators -- Security Guards and Gaming Surveillance Officers -- Licensure -- Employment -- Executive Protection Services -- Public Police and Private Security -- Chapter Summary -- Key Terms -- Discussion Questions -- Internet Exercises -- pt. V LOOKING AHEAD -- ch. 16 The Future of Policing in America -- Chapter Learning Objectives -- The Changing Police Role -- Research and Planning as Police Functions -- Changing the Police Image -- Community and Problem-Oriented Policing -- Note continued: Civilianization -- Accreditation -- Lateral Entry -- Training -- Education -- Police Leadership -- Private and Contract Security Personnel -- Technological Changes in Policing -- Clarifying the Police Role -- Predictive Policing -- The Role of Community Policing -- Intelligence-Led Policing and Terrorism -- Terrorism Laws -- Chapter Summary -- Key Terms -- Discussion Questions -- Internet Exercises. Subjects: Police--United States. Community policing--United States. Notes: Scaramella's name appears first on the earlier edition. Includes bibliographical references and index. Dewey class no.: 363.20973

Justice denied: the role of forensic science in the miscarriage of justice LCCN: 2014426003 Type of material: Book Personal name: Klatzow, David. Main title: Justice denied: the role of forensic science in the miscarriage of justice / David Klatzow. Published/Produced: Cape Town, South Africa: Zebra Press, 2014. Description: ix, 274 pages, 8 unnumbered pages of plates: illustrations (some color); 23 cm ISBN: 9781770226944 9781770226968 (PDF) LC classification: HV8073.K528 2014 Subjects: Forensic sciences--Case studies. Forensic sciences--South Africa--Case studies. Criminal investigation--Case studies. Criminal investigation--South Africa--Case studies. Criminal justice, Administration of--Case studies. Criminal justice, Administration of--South Africa. Notes: Includes bibliographical references (pages 251-263) and index.

Law, rules and procedures of criminal investigation in Nigeria LCCN: 2014380407 Type of material: Book Personal name: Momodu, Basil, author. Main title: Law, rules and procedures of criminal investigation in Nigeria / Basil Momodu. Published/Produced: [Benin City, Nigcria]: Evcrgrccn Overseas Publications Limited, 2013. Description: xxviii, 630 pages; 24 cm ISBN: 9789789082070 978908207X LC classification: KTA4636.M656 2013 Related names: Nigeria. Evidence Act, 2011. Contents: Investigative field notes, photographs, sketches and forensic evidence -- Experts in forensic science and the admissibility of forensic evidence in court -- Scientific aid to criminal investigation -- Crime scene investigation -- Effecting an arrest: procedure and power -- Use of handcuffs and power to detain -- Searching for evidence -- Questioning of suspect: procedure and legal limits (part 1) -- Questioning of suspect: procedure and legal limits (part 2) -- Use of force: legal limits -- Homicide investigation -- Investigation of rape -- Investigation of robbery -- Investigating questioned documents -- Obtaining physical characteristics of suspects: the legal limits -- Transnational crimes investigation -- Investigative reports -- Case diary: contents, arrangement and use -- Testifying in court -- Charges, arraignment and proof of crime -- Witnesses -- Defences to criminal liability -- Police investigations and the rights of suspects -- The investigator and the assignment of criminal investigation -- Police: the limits of discretionary power -- Money laundering of international dimension: an investigative guide -- Provention and detection of fraud -- Making fraud investigation easy in the United States of America -- Suspect's rights and other issues relating to identification parade -- Appendix 1: [sample cases] -- Appendix 2: Mutual Assistance in Criminal Matters within the Commonwealth (Enactment and Enforcement) Act, Cap. M24 [1998] -- Evidence Act, 2011. Subjects: Criminal investigation--Nigeria. Notes: Includes bibliographical references and index. Dewey class no.: 363.2509669

Mark & trace analysis LCCN: 2013006962 Type of material: Book Personal name: Hunter, William, 1971- Main title: Mark & trace analysis / William Hunter. Published/Created: Philadelphia, Pa.: Mason Crest Publishers, c2014. Description: xii, 253 pages; 24 cm. ISBN: 9781422228715 (hardcover) 9781422228616 (series) 9781422289570 (ebook) LC classification: HV8073.8.H789 2014 Variant title: Mark and trace analysis Contents: Introducing forensic science! -- Physical evidence -- Trace evidence that leads to convictions -- Cracking cases with tool marks -- Working with mark and trace evidence. Subjects: Forensic sciences--Juvenile literature. Trace analysis--Juvenile literature. Trace evidence--Juvenile literature. Evidence, Criminal--Juvenile literature. Criminal investigation--Juvenile literature. Notes: Includes bibliographical references and index. 012. Grades 7 to 8. Series: Solving crimes with science, forensics Dewey class no.: 363.25/62

Reconceptualizing the literacies in adolescents' lives: bridging the everyday/academic divide LCCN: 2011026130 Type of material: Book Main title: Reconceptualizing the literacies in adolescents' lives: bridging the everyday/academic divide / edited by Donna E. Alvermann, Kathleen A. Hinchman. Edition: 3rd ed. Published/Created: New York: Routledge, 2012. Description: xvi, 279 p.: ill.; 24 cm. ISBN: 9780415892919 (alk. paper) 0415892910 (alk. paper) 9780415892926 (pbk.: alk. paper) 0415892929 (pbk.: alk. paper) 9780203817285 (ebk) 0203817281 (ebk) LC classification: LB1631.R296 2012 Related names: Alvermann, Donna E.

Hinchman, Kathleen A. Contents: Machine generated contents note: pt. I Understanding Youth's Everyday Literacies -- 1.Touchstone Chapter Playing for Real: Texts and the Performance of Identity / Lorri Neilsen Glenn -- 2.Becoming Life-Long Readers: Insights from a Comic Book Reader / Stergios G. Botzakis -- 3.Low-Income Youth's (Public) Internet Practices in South America: Potential Lessons for Educators in the U.S. and Other Post-Industrial Nations / Eliane Rubinstein-Avila -- 4.Teacher Response: Lessons Learned from Young People's Everyday Literacies / Margaret Moran -- pt. II Integrating Everyday and Academic Literacies -- 5.Touchstone Chapter "Struggling" Adolescents' Engagement in Multimediating: Countering the Institutional Construction of Incompetence / David O'Brien -- 6.Thinking with Forensic Science: A Content Analysis of Forensic Comic Books and Graphic Novels / Marcia Mardis -- 7.Reclaiming and Rebuilding the Writer Identities of Black Adolescent Males / Marcelle M. Haddix -- 8.Teacher Response: Bridging Everyday Literacies with Academic Literacy / McKenzie Weaver -- pt. III Addressing Sociocultural and Identity Issues in Adolescents' Literacy Lives -- 9.Touchstone Chapter Exploring Race, Language, and Culture in Critical Literacy Classrooms / Renee Moore -- 10.Re-Writing the Stock Stories of Urban Adolescents: Autobiography as a Social and Performative Practice at the Intersections of Identities / Lalitha Vasudevan -- 11."In This Little Town Nothing Much Ever Happens, But Someday Something Will": Reading Young Adult Literature from the Blue Ridge Foothills / Gay Ivey -- 12.Teacher Response: Addressing Sociocultural and Identity Issues in Adolescents' Literacy Lives / George White -- pt. IV Changing Teachers, Teaching Changes -- 13.Touchstone Chapter Adolescents' Multiple Identities and Teacher Professional Development / Alfred W. Tatum -- 14.Reconceptualizing Together: Exploring Participatory and Productive Critical Media Literacies in a Collaborative Teacher Research Group / Ellen Damlich -- 15.Middle School Teachers' Successes and Challenges in Navigating Web 2.0 Technologies in a Web 1.0 Middle School / Margaret C.

Hagood -- 16.Teacher Response: Professional Development to Reconceptualize Literacy Instruction / Elizabeth G. Mascia. Subjects: Language arts (Secondary)--Social aspects--United States. Literacy--Social aspects--United States. Critical pedagogy--United States. Notes: Includes bibliographical references and index. Dewey class no.: 428.0071/2

Scientific protocols for fire investigation LCCN: 2012015769 Type of material: Book Personal name: Lentini, John J. Main title: Scientific protocols for fire investigation / John J. Lentini. Edition: 2nd ed. Published/Created: Boca Raton, FL: Taylor & Francis, c2013. Description: 624p.: col. ill.; 26 cm. Links: Cover image http://jacketsearch.tandf.co.uk/common/jackets/covers/websmall/978143987/9781439875988.jpg ISBN: 9781439875988 (hardback) LC classification: TH9180.L46 2013 Summary: "Providing the critical information necessary to conduct an effective fire investigation, this new edition of a best-selling text includes four new cases and new and expanded discussion of power strips and surge suppressors, space heater fires, source identification, and changes in NFP A 921. Integrating historical, theoretical, and practical aspects in a scientific context, specifically chemistry, physics, and fluid dynamics, it explains the latest lab techniques, ethics codes, mathematical prediction, and forensic reports as well as new graphics showing predicted versus measured phenomena"--Provided by publisher. Subjects: Fire investigation. Notes: Includes bibliographical references and index. Series: Protocols in forensic science series Dewey class no.: 363.37/65

Silent witnesses: the often gruesome but always fascinating history of forensic science LCCN: 2014006713 Type of material: Book Personal name: McCrery, Nigel, 1953- Main title: Silent witnesses: the often gruesome but always fascinating history of forensic science / Nigel McCrery. Published/Produced: Chicago, Illinois: Chicago Review Press, 2014. Description: xxiii, 264 pages: illustrations (some color); 24 cm ISBN: 9781613730027 (paperback) 1613730020 (paperback) LC classification: HV8073.M3323 2014 Summary: "Crime novelist and former police officer Nigel McCrery provides an account of all the major areas of forensic science from around the world over the past two centuries. The book weaves dramatic narrative and scientific principles together in a way that allows readers to figure out crimes along with the experts. Readers are introduced to such fascinating figures as Dr. Edmond Locard, the "French Sherlock Holmes;" Edward Heinrich, "Wizard of Berkeley," who is credited with having solved more than 2,000 crimes; and Alphonse Bertillon, the French scientist whose guiding principle, "no two individuals share the same characteristics," became the core of criminal identification. Landmark crime investigations examined in depth include a notorious murder involving blood evidence and defended by F. Lee Bailey, the seminal 1936 murder that demonstrated the usefulness of the microscope in examining trace evidence, the 1849 murder of a wealthy Boston businessman that demonstrated how difficult it is to successfully dispose of a corpse, and many others"-- Provided by publisher. "Through examinations of specific cases throughout history, crime novelist and former police officer Nigel McCrery explores the gruesome but fascinating history and progress of forensic science around the world"-- Provided by publisher. Subjects: Forensic sciences--History. Forensic sciences--Case studies. Criminal investigation--Case studies. SCIENCE / General. MEDICAL / Forensic Medicine. TRUE CRIME / General. Notes: "First published in Great Britain in 2013 by Random House Books." Includes index. Other edition: Reproduction of (manifestation): McCrery, Nigel, 1953- Silent witnesses London: Random House, 2013 (OCoLC) 828418415 Dewey class no.: 363.2509

Simpson's forensic medicine LCCN: 2013048944 Type of material: Book Personal name: McGovern, Cliona, author. Main title: Simpson's forensic medicine / Cliona McGovern, Jason Payne-James, Steven B. Karch, Richard Jones, John Manlove. Edition: [13th edition] Irish edition. Published/Produced: Boca Raton: Taylor & Francis, 2014. Description: p.; cm. ISBN: 9781482226126 (pbk.: acid-free paper) 148222612X (pbk.: acid-free paper) LC classification: RA1051 Variant title: Forensic medicine Related names: Payne-

James, Jason, author. Karch, Steven B., author. Jones, Richard (Forensic pathologist), author. Manlove, John (Forensic scientist), author. Contents: Principles of forensic practice -- The ethics of medical practice -- The medical aspects of death -- Identification of the living and the dead -- The appearance of the body after death -- Unexpected and sudden death from natural causes -- Deaths and injury in infancy -- Assessment, classification and documentation of injury -- Regional injuries -- Ballistic injuries -- Use of force and restraint -- Sexual assault -- Child assault and protection -- Transportation medicine -- Asphyxia -- Immersion and drowning -- Heat, cold and electrical trauma -- Principles of toxicology -- Alcohol -- Licit and illicit drugs -- Medicinal poisons -- Miscellaneous poisons -- Principles of forensic science -- Allied forensic specialties. Subjects: Forensic Medicine. Notes: Preceded by Simpson's forensic medicine / Jason Payne-James... [et al.]. 13th ed. c2011. Includes bibliographical references and index. Dewey class no.: 614/.1

So, you want to work with the ancient and recent dead?: unearthing careers from paleontology to forensic science LCCN: 2014047433 Type of material: Book Personal name: Bedell, J. M. (Jane M.), author. Main title: So, you want to work with the ancient and recent dead?: unearthing careers from paleontology to forensic science / by J. M. Bedell. Published/Produced: New York: Aladdin; Hillsboro, Oregon: Beyond Words, [2015] Projected pub date: 1510 Description: pages cm. ISBN: 9781582705460 (hardcover: alk. paper) 9781582705453 (pbk.: alk. paper) LC classification: HV8073.8.B44 2015 Contents: Choosing a career working with the dead -- Discovering the ancient dead: humans and civilizations -- Learning about the ancient dead: animals and plants -- Caring for the recent dead -- Preserving the dead -- Studying the dead -- Remembering the dead -- Embracing the dead in the arts -- Putting the dead to work -- Research, extinction, and other dead things -- Industry resources. Subjects: Forensic sciences--Vocational guidance--Juvenile literature. Forensic anthropology--Vocational guidance--Juvenile literature. Archaeology--Vocational guidance--Juvenile literature.

Paleontology--Vocational guidance--Juvenile literature. Series: Be what you want series Be what you want series. Dewey class no.: 331.702

Solving crimes with physics LCCN: 2013006971 Type of material: Book Personal name: Hunter, William, 1971- Main title: Solving crimes with physics / William Hunter. Published/Created: Broomall, Pa.: Mason Crest Publishers, c2014. Description: 112 p.: col. ill.; 25 cm. ISBN: 9781422228739 (hardcover) 1422228738 (hardcover) 9781422228616 (series) 1422228614 (series) LC classification: HV8073.H8942 2014 Contents: Introducing forensic science -- Introducing physics -- All about blood -- Blood spatter -- Using physics: rifling and ballistics -- Using physics: bombs and explosions -- Focus on career: becoming a forensic physics expert. Subjects: Evidence, Criminal--Juvenile literature. Criminal investigation--Juvenile literature. Forensic sciences--Juvenile literature. Notes: Includes bibliographical references (p. 107-108) and index. 012. Grades 7 to 8. Series: Solving crimes with science, forensics Dewey class no.: 363.25

Statistical analysis in forensic science: evidential value of multivariate physicochemical data LCCN: 2013031698 Type of material: Book Main title: Statistical analysis in forensic science: evidential value of multivariate physicochemical data / Grzegorz Zadora, Agnieszka Martyna, Daniel Ramos, Colin Aitken. Published/Created: Chichester, West Sussex: John Wiley & Sons Inc., c2014. Description: xiv, 322 p.: ill.; 25 cm. ISBN: 9780470972106 (cloth) LC classification: RA1057.S73 2014 Related names: Zadora, Grzegorz. Martyna, Agnieszka. Ramos, Daniel. Aitken, Colin. Subjects: Chemistry, Forensic. Forensic statistics. Chemometrics. Notes: Includes bibliographical references and index. Additional formats: Online version: Statistical analysis in forensic science Chichester, West Sussex: John Wiley & Sons Inc., c2014 9781118763179 (DLC) 2013038429 Dewey class no.: 614/.12

Textbook of diagnostic microbiology LCCN: 2013045846 Type of material: Book Main title:

Textbook of diagnostic microbiology / [edited by] Connie R. Mahon, MS, Director, Staff and Organization Development, Health Resources and Services Administration, HIV/AIDS Bureau, Rockville, Maryland, Adjunct Faculty, Department of Clinical Research and Leadership, School of Medicine and Health Sciences, George Washington University, Washington, DC, Donald C. Lehman, EdD, MT(ASCP), SM(NRM), Associate Professor, Department of Medical Laboratory Sciences, University of Delaware, Newark, Delaware, George Manuselis, MA, MT(ASCP), Emeritus, Medical Technology Division, Ohio State University, Columbus, Ohio, Adjunct Faculty, Department of Natural Sciences and Forensic Science, Central Ohio Technical College, Newark, Ohio. Edition: Fifth edition. Published/Produced: Maryland Heights, Missouri: Elsevier, [2015] Description: xvii, 1076 pages: illustrations (chiefly color); 29 cm ISBN: 9780323089890 (hardcover) LC classification: QR67.T49 2015 Related names: Mahon, Connie R., editor of compilation. Lehman, Donald C., editor of compilation. Manuselis, George, editor of compilation. Subjects: Microbiological Techniques. Bacterial Infections--diagnosis. Communicable Diseases--diagnosis. Mycoses--diagnosis. Virus Diseases--diagnosis. Notes: Includes bibliographical references and index. Dewey class no.: 616.9/041

The global practice of forensic science LCCN: 2014035953 Type of material: Book Main title: The global practice of forensic science / Douglas H. Ubelaker, editor. Published/Produced: Chichester, West Sussex, UK; Hoboken, NJ: John Wiley & Sons Inc., [2015] Projected pub date: 1503 Description: pages cm. ISBN: 9781118724163 (hardback) LC classification: HV8073.G568 2015 Related names: Ubelaker, Douglas H. Summary: "The Global Practice of Forensic Science presents histories, issues, patterns, and diversity in the applications of international forensic science. Written by 64 experienced and internationally recognized forensic scientists, the volume documents the practice of forensic science in 28 countries from Africa, the Americas, Asia, Australia and Europe. Each country's chapter explores factors of political history, academic linkages, the influence of individual cases, facility development, types of cases examined, integration within forensic science, recruitment, training, funding, certification, accreditation, quality control, technology, disaster preparedness, legal issues, research and future directions. Aimed at all scholars interested in international forensic science, the volume provides detail on the diverse fields within forensic science and their applications around the world"-- Provided by publisher. "The Global Practice of Forensic Science examines the practice of forensic science between around 30 countries from across five continents"-- Provided by publisher. Subjects: Forensic sciences. LAW / Forensic Science. Notes: Includes index. Additional formats: Online version: Global practice of forensic science Chichester, West Sussex, UK; Hoboken, NJ: John Wiley & Sons Inc., [2015] 9781118724224 (DLC) 2014037173 Series: Forensic science in focus Dewey class no.: 363.

The inheritor's powder: a tale of arsenic, murder, and the new forensic science LCCN: 2013024989 Type of material: Book Personal name: Hempel, Sandra. Main title: The inheritor's powder: a tale of arsenic, murder, and the new forensic science / Sandra Hempel. Edition: First American edition. Published/Produced: New York: W. W. Norton & Company, Inc., 2013. Description: viii, 278 pages; 25 cm ISBN: 9780393239713 (hardcover) LC classification: HV6555.G7 H46 2013 Subjects: Poisoning--Great Britain--History--19th century. Arsenic--Toxicology--Great Britain--History--19th century. Murder--Great Britain--History--19th century. Toxicology--Great Britain--History--19th century. Forensic toxicology Notes: Includes bibliographical references and index. Dewey class no.: 364.152/3092

The Kelly Gang unmasked LCCN: 2013387378 Type of material: Book Personal name: MacFarlane, Ian, author. Main title: The Kelly Gang unmasked / Ian MacFarlane. Published/Produced: South Melbourne, Victoria: Oxford University Press, 2012. Description: xii, 258 pages, 16 pages of plates: illustrations, 1 map; 23 cm ISBN: 9780195519662 (pbk.) 0195519663 (pbk.) Summary: This book re-examines the story of the Kelly Gang in fascinating detail and with many new insights. The mythology created by pro-Kelly writers is critically explored, unravelled, and often found wanting. Many missing official documents have been identified for the first time. Contents: "I am bullet proof. You can't hurt me" -- Organised rural crime and terror -- The killing ground at Stringybark Creek -- Forensic science investigations -- "We shall all have to go to a bigger court than this" -- Victoria police actions defended -- The bitter aftertaste of the Kelly legend. Subjects: Kelly, Ned, 1855-1880. Bushrangers--Australia--Victoria--History--19th century. Criminals--Australia--Victoria--History--19th century. Violent crimes--Australia--Victoria--History--19th century. Police--Violence against--Australia--Victoria--History--19th century. Notes: Includes bibliographical references (pages 245-246) and index. Dewey class no.: 364.15509945

The law of science and the science of law: cases in forensic science LCCN: 2009045672 Type of material: Book Personal name: Myers, Bradley. Main title: The law of science and the science of law: cases in forensic science / Bradley K. Myers, A. Burton Bass, and Hyman D. Gesser. Published/Created: New York: Nova Science Publishers, c2012. Description: 330 p.: ill.; 26 cm. ISBN: 9781608768073 (hardcover) LC classification: K5485.M94 2012 Related names: Bass, A. B. Gesser, Hyman D., 1929- Contents: Evidence and the expert witness -- The breath analysis for alcohol and problems with error -- The poison pill -- House fire #1 -- House fire # 2 -- Slip and fall -- The phenol spill -- Thunderstorm flooding or hedge your bets -- Expert liability -- Summing up, conclusions & appraisals. Subjects: Chemistry, Forensic. Evidence, Expert--United States. Evidence, Expert--Canada. Notes: Includes bibliographical references and index. Series: Laws and legislation Dewey class no.: 363.25

The neuroscience of handwriting: applications for forensic document examination LCCN: 2011038465 Type of material: Book Personal name: Caligiuri, Michael P. Main title: The neuroscience of handwriting: applications for forensic document examination / Michael P. Caligiuri and Linton Mohammed. Published/Created: Boca Raton: Taylor & Francis,

2012. Projected pub date: 1203 Description: p. cm. ISBN: 9781439871409 (alk. paper) LC classification: HV8074.C25 2012 Related names: Mohammed, Linton. Contents: Neuroanatomy and neurochemistry of hand motor control -- Neuroanatomical basis of handwriting -- Models of motor control -- Disease and motor control -- Psychotropic medication, substance abuse, and motor control -- Aging and motor control -- A kinematic approach to signature authentication -- Isochrony in genuine and auto-simulated signatures -- Kinematic variability and stroke direction -- Examination of signatures and handwriting -- Neurologic disease and handwriting -- Psychotropic medication and handwriting -- Substance abuse and handwriting -- Aging and handwriting -- Conclusions. Subjects: Writing--Identification. Signatures (Writing) Legal documents--Identification. Graphology. Notes: Includes bibliographical references and index. Series: International forensic science and investigation series; 25 Dewey class no.: 363.25/65.

The Routledge international handbook on hate crime LCCN: 2013047731 Type of material: Book Main title: The Routledge international handbook on hate crime / edited by Nathan Hall, Abbee Corb, Paul Giannasi and John G. D. Grieve; foreword by Neville Lawrence, OBE. Edition: First Edition. Published/Produced: New York: Routledge, Taylor & Francis Group, 2015. Description: xviii, 440 pages; 26 cm ISBN: 9780415818902 (hardback: alk. paper) LC classification: HV8079.H38 R68 2015 Variant title: International handbook on hate crime Related names: Hall, Nathan, editor of compilation. Contents: Introduction / Nathan Hall, Abbee Corb, Paul Giannasi, and John Grieve -- Theories and concepts -- Framing the boundaries of hate crime / Neil Chakraborti -- Beyond the silo: rethinking hate crime and intersectionality / Hannah Mason-Bish -- The personal injuries of hate crime / Paul Iganski and Spiridoula Lagou -- Exploring the community impacts of hate crime / Barbara Perry -- Legislating against hate / Gail Mason -- Explaining hate crimes: sociological and criminological perspectives / Nathan Hall -- Explaining hate crimes: perspectives from the wider social sciences / Nathan

Hall -- The international geography of hate -- Hate crimes in Europe / Mike Whine -- Hate crimes in the UK / Paul Giannasi -- Sectarianism and hate crime in Northern Ireland / Marian Duggan -- Global antisemitism / Dave Rich -- The European extreme right / Emmanuel Godin -- Hate crimes in the United States / Jordan Blair Woods -- Hate crimes in Canada / Abbee Corb -- A governance of denial: hate crime in Australia and New Zealand / Nicole Asquith -- Key issues in hate crime -- Hate crimes against disabled people / Chih Hoong Sin -- Disability hostility, harassment and violence in the UK / Paul Hamilton and Loretta Trickett -- Alternative subcultures and hate crime / Jon Garland and Paul Hodkinson -- Hate crimes against gypsies, travellers and Roma in Europe / Zoe James -- Reflections on gendered masculine identities in targeted violence against ethnic minorities / Loretta Trickett -- LGBT hate crime / Leslie Moran -- Anti-transgender hate crimes Jordan / Blair Woods and Jody L. Herman -- Good and evil on the internet / Sol Littman -- Online hatred / Sarah Rohlfing -- Online hate and cyber-bigotry: a glance at our radicalized online world Abbee Corb -- Hate crimes in sport / Nick Hawkins -- Combating hate and hate crime -- Policing and hate crime / Paul Giannasi -- Intelligence-led approaches to combating hate crime / John Grieve -- Forensic science and hate crime / Paul Smith -- You're a victim, don't become an offender: a study of the "moral career" of racist hate crime victims / Corinne Funnell -- Working with offenders / Liz Dixon and David Court -- Rehabilitative programmes for hate offenders / Eila Davis -- Repairing the harms of hate crime: a restorative justice approach / Mark Walters -- Challenging sectarianism / Graham Spencer -- Deradicalisation / Daniel Koehler. Subjects: Hate crimes--Handbooks, manuals, etc. Notes: Includes bibliographical references and index. Dewey class no.: 364.15

Transnational television drama: special relations and mutual influence between the US and UK LCCN: 2012462669 Type of material: Book Personal name: Weissmann, Elke. Main title: Transnational television drama: special relations and mutual influence between the US and UK / Elke Weissmann. Published/Created: Houndmills, Basingstoke, Hampshire; New

York: Palgrave Macmillan, 2012. Description: x, 225 p.: ill.; 23 cm. Links: Contributor biographical information http://www.loc.gov/catdir/enhancements/fy1403/2012462669-b.html Publisher description http://www.loc.gov/catdir/enhancements/fy1403/2012462669-d.html Table of contents only http://www.loc.gov/catdir/enhancements/fy1409/2012462669-t.html ISBN: 9780230297753 0230297757 LC classification: PN1992.3.G7 W45 2012 Summary: A history of American and British television drama, this book charts how the two production systems have moved closer together since the 1970s: both observe each other to drive innovation, and both continuously turn to each other to find new markets and new production partners. Although earlier collaborations exist, this increased transnationalisation of US and UK television drama has intensified since the 1970s as the increased number of channels and new technologies such as the internet and cable and satellite have led to stronger competition. Examining genres as diverse as period drama, the mini-series, the super-soaps of Dallas and Dynasty, crime drama and the recent spout of celebrated British and American quality drama, this book investigates how marketing campaigns within the press continually return these dramas into the realm of the nations they represent. Contents: US and UK television, 1970 to 2010 -- National difference and transnational assimilation -- Transnational aesthetic influences and performance -- The development of a new genre -- Transnational audiences of forensic science drama -- The national in international co-productions -- National fantasies of 'American quality drama.' Subjects: Television series--Great Britain. Television series--United States. Television and globalization. Television broadcasting--International cooperation. Notes: Includes bibliographical references (p. 195-215) and index. Dewey class no.: 791.450941

Wildlife DNA analysis: applications in forensic science LCCN: 2012050747 Type of material: Book Personal name: Linacre, Adrian. Main title: Wildlife DNA analysis: applications in forensic science / Adrian M.T. Linacre and Shanan S. Tobe. Published/Produced: Chichester, West Sussex, UK: John Wiley & Sons Inc., 2013. Description: xiv, 334 pages; 25 cm ISBN:

9780470665954 (hardback) 9780470665961 (pbk) LC classification: HV8079.W58 L56 2013 Related names: Tobe, Shanan S. Summary: "This book is the first to approach the fast developing field of wildlife forensics with a focus on the application of DNA profiling and analysis. Case studies throughout link theory and practice and highlight the use of DNA testing in species testing. The text assumes only a basic background knowledge of DNA, so offers information boxes clarifying technical information, step-by-step guidance on sequence comparisons, and a discussion of the different markers used in species testing. This produces a highly accessible introduction for both students and forensic professionals"-- Provided by publisher. "Wildlife DNA Analysis: Applications in Forensic Science provides an accessible introduction to the application of DNA typing in wildlife forensic science"-- Provided by publisher. Subjects: Wildlife crime investigation. Forensic sciences. MEDICAL / Forensic Medicine. Notes: Includes bibliographical references and index. Additional formats: Online version: Linacre, Adrian. Wildlife DNA analysis Chichester, West Sussex, United Kingdom: John Wiley & Sons Inc., 2013 9781118496527 (DLC) 2013007454 Dewey class no.: 363.25/9628591

Working in the killing fields: forensic science in Bosnia LCCN: 2014044972 Type of material: Book Personal name: Ball, Howard, 1937- Main title: Working in the killing fields: forensic science in Bosnia / Howard Ball. Published/Produced: Lincoln: Potomac Books, [2015] Projected pub date: 1505 Description: pages cm ISBN: 9781612347189 (cloth: alk. paper) LC classification: DR1313.7.A85 B354 2015 Subjects: Genocide--Bosnia and Hercegovina. Yugoslav War, 1991-1995--Bosnia and Hercegovina. Dead--Identification. Disappeared persons--Bosnia and Hercegovina. Forensic sciences--Bosnia and Hercegovina. Bosnia and Herzegovina--History--1992- Notes: Includes bibliographical references and index. Dewey class no.: 949.703

INDEX

A

B

C

F

G

H

I

J

K

L

M

N

O

P

Q

R

S

T